48 Structure and Bonding

Biochemistry

With Contributions by
I. Bertini P. Hemmerich J. Livorness
C. Luchinat V. Massey H. Michel
C. Schug A. Scozzafava T. D. Smith

With 46 Figures and 14 Tables

Springer-Verlag
Berlin Heidelberg New York 1982

ISBN 3-540-10986-2 Springer-Verlag Berlin Heidelberg New York
ISBN 0-387-10986-2 Springer Verlag New York Heidelberg Berlin

Library of Congress Catalog Card Number 67-11280

Typesetting and printing: Schwetzinger Verlagsdruckerei. Bookbinding: J. Schäffer, Grünstadt.
2152/3140-543210

Peter Hemmerich
30. 12. 1929 – 3. 10. 1981

Table of Contents

The Role of Manganese in Photosynthesis

John Livorness and Thomas D. Smith

Chemistry Department, Monash University, Clayton, Victoria, 3168, Australia

Although the process of light energy conversion to chemical form by specialized plant cell organelles, the chloroplasts, is well understood, many contemporary investigations have been concerned with the catalytic role played by manganese in oxygen evolution. Current views of the photosynthetic electron transport chain or the so-called "Z-scheme" intermediates identify the cytochromes b_6, $b\text{-}_{559}$ and f, the chlorophyll linked primary electron acceptor and donors of photosystems -I and -II and a number of e.p.r. distinct iron-sulfur centres involved in non cyclic and possibly cyclic electron transfer pathways. The accummulation of four redox equivalents leads to the oxidation of water to yield molecular oxygen by a water splitting metallo-enzyme (designated -Y) which involves at least two manganese atoms per molecule of enzyme. During the four electron oxidation of water, e.p.r. and proton relaxation evidence suggest the two manganese atoms undergo redox changes between manganese(II) and manganese(III) states. A number of proposed manganese model systems exhibiting the features required by the well known S-state mechanism of oxygen evolution are discussed. The modelling of both the charge separation function of the photosystems and the properties of the catalytically active manganese is of current interest from an energy production point of view.

Structure and Bonding 48

1 Introduction

Photosynthesis may be defined as the assimilation of carbon dioxide in the light to form carbohydrates. The chemical identity of the electron donor is water in all photosynthetic organisms except bacteria. The overall equation of photosynthesis for this group of organisms is:

$$H_2O + CO_2 \xrightarrow{\text{light}} (CH_2O) + O_2 \ .$$

Photosynthetic bacteria usually neither produce nor use molecular oxygen, most of them are strict anaerobes poisoned by oxygen. Instead of water these organisms use other compounds as electron donors. The green and purple sulphur bacteria use hydrogen sulphide:

$$2\,H_2S + CO_2 \xrightarrow{\text{light}} (CH_2O) + H_2O + 2\,S \ .$$

Some non-sulphur purple bacteria use an organic hydrogen donor, such as isopropanol, which is oxidized to acetone:

$$2\,CH_3CHOHCH_3 + CO_2 \xrightarrow{\text{light}} (CH_2O) + 2\,CH_3COCH_3 + H_2O \ .$$

In addition to alternative electron donors, carbon dioxide is not the only electron acceptor in general photosynthesis; NO_3^- or N_2 and H^+ may be reduced to NH_3 and H_2 respectively.

The magnitude and role of photosynthesis is illustrated by considering the yearly turnover of carbon fixation by plants, about 2×10^{11} tonnes of carbon with an energy content of 3×10^{12} J or about $10 \times$ the worlds annual energy use and $200 \times$ its food energy comsumption. All the atmospheric carbon dioxide is cycled through plants every 300 years, all the oxygen every 2000 years and all the water every 2×10^6 years. The utilization of this enormous capacity of hydrocarbon and possibly ammonia and hydrogen production from a free energy source, the sun, has attracted much interest in the light of impending shortages of fossil-fuel based energy supplies [1–3].

It has now been established that in plant photosynthesis light driven electron transport, coupled to phosphorylation, produces oxygen, NADPH (reduced nicotinamide adenine dinucleotide phosphate) and ATP (adenosine triphosphate). The latter two are in turn consumed in, and drive the CO_2-reduction-cycle. This cycle is composed of a series of light independant reactions and has been extensively studied [4,5].

The trapping and conversion of light energy in higher plants and green algae occurs in highly organized ellipsoid structures called chloroplasts which float free in the cytoplasm. Electron microscopic pictures show that chloroplasts in higher plants are saucer-shaped bodies 4–10 μm in diameter and 1 μm in thickness with an outer membrane or envelope separating it from the rest of the cytoplasm.

Internally the chloroplast comprises a system of lamellae or flattened thylakoids arranged in stacks in certain dense green regions, termed the grana (Fig. 1). Each

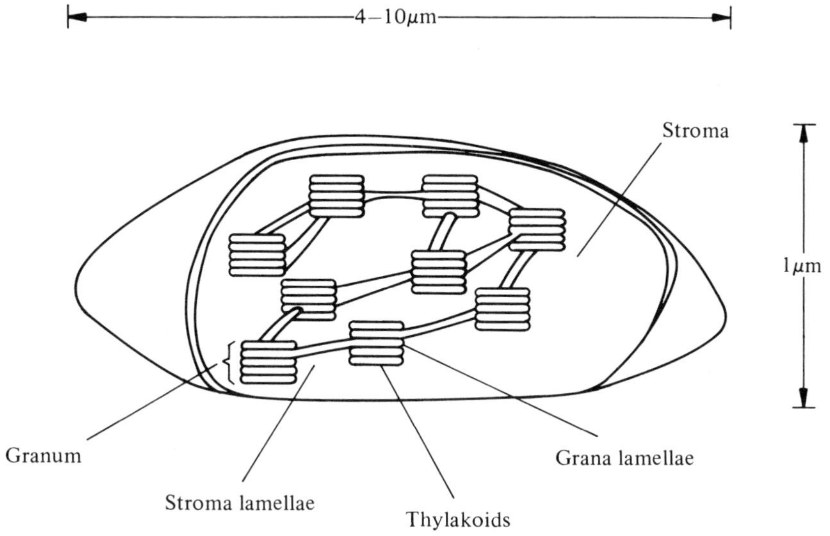

Fig. 1. Chloroplast organization

lamellae in the chloroplast may contain two double-layer membranes. The grana are embedded in a colourless matrix, the stroma, and the whole chloroplast is surrounded by a boundary of double lipid-protein membrane[6]. Embedded within the thylakoid membrane is the photochemical apparatus which consists of pigments (chlorophylls and carotenoids) electron carriers and the enzymes responsible for ATP, NADPH and oxygen generation. Other proteins (ribulose diphosphate carboxy-dismutase) are also associated with the thylakoid structure. A more detailed description of the chloroplast ultrastructure can be found in recent reviews[7–9].

2 Photosynthetic Electron Transport

The inner membrane systems of chloroplasts, the thylakoids, transforms energy of light quanta into other forms of energy which are used to drive the secondary processes of photosynthesis. Fully matured spinach thylakoids have an estimated chlorophyll content of $> 10^5$ molecules. Two functionally distinct pools of chlorophyll have been identified, a) antennae or light harvesting chlorophyll (also containing accessory pigments such as the carotenoids), and b) energy trapping and converting chlorophyll-a (Fig. 2). The former pool absorbs visible light energy which is rapidly transferred to the smaller pools of specialized chlorophyll which form the primary electron donors of the two photosystems, PSI and PSII, of the electron transport chain as depicted by Fig. 3. The two pools are designated P680 (PSII reaction centre) and P700 (PSI reaction centre) and are maximally activated by light of wavelengths

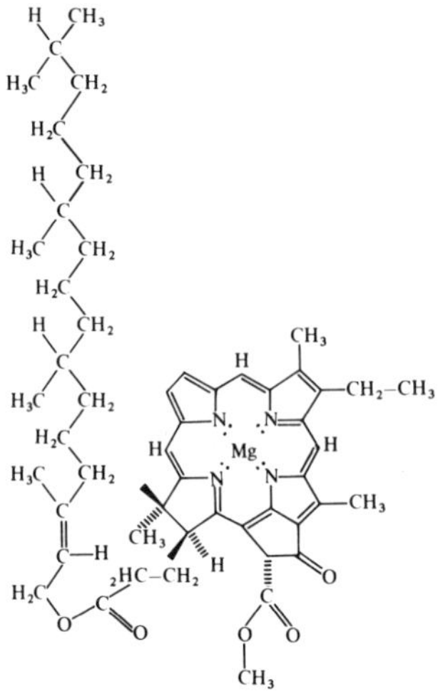

Fig. 2. Structure of chlorophyll-a

around 680 nm and 700 nm respectively. The antennae pool of P680 and P700 consist of 300 chlorophyll molecules per reaction centre. The reaction centres P680 and P700 communicate at antennae level through a low molecular weight chlorophyll protein (M.W. 25000 Daltons) designated, LH, light harvesting protein [10] which is though to act as a functional component which can transfer excitation energy to both PSI and PSII and can participate in energy transfer from PSII to PSI [11–14]. The regulation of excitation energy distribution is thought to be mediated by magnesium cations [15] although some doubt on the function of magnesium has recently been expressed [16].

The structure, function and organization of chlorophyll *in vivo* has been extensively reviewed [17–20]. Chlorophyll aggregates are thought to comprise the antennae and photoreaction centres P680, P700 and P865 in the case of photosynthetic bacteria (the chlorophyll associated with bacteria is distinct from plant chlorophyll and is termed Bacteriochlorophyll, BChl.). *In vitro* modelling of chlorophyll aggregates has been the focus of extensive study [17,18,22,25–27]. P680 and P700 are thought to be hydrated chlorophyll-a dimers or special pairs which are closely related to a protein component which plays an important role in controlling the rate of electron transfer and maximizing the efficiency of the primary reaction [28]. Recent optical and electron paramagnetic resonance studies of oxidized magnesium tetraphenylchlorin, a synthetic model for chlorophyll, suggest that, by comparison with P680 *in vivo*, P680 may in fact be a ligated monomer whose function as a phototrap is determined by interactions possibly with its protein environment [29].

The excitation of P680 and P700 gives rise to oxidized, e.p.r. detectable [30–32] free radical chlorophyll species $P700^{+\bullet}$ and possibly $P680^{+\bullet}$. $P700^{+\bullet}$ exhibits a prominent reversible component at g = 2.0025 showing a Gaussian line shape with a peak-to-peak line width of about 7.0 G [32–34]. Optical measurements of the decrease in light absorption (at 700 nm) or photobleaching are consistent with the e.p.r. findings which suggest the formation of cationic chlorophyll free radicals derived from photo induced electron ejection. A signal with g = 2.002 and width about 8 G showing decay

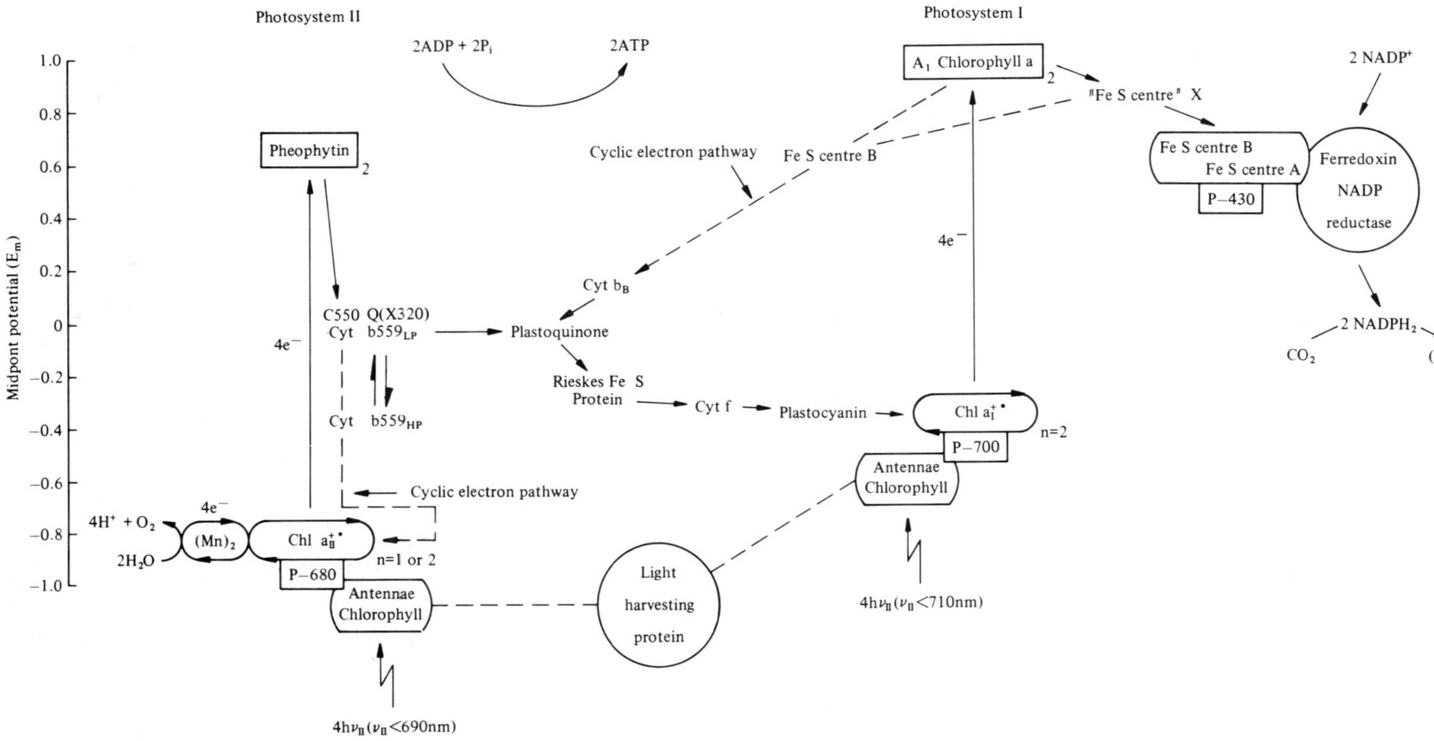

Fig. 3. Electron transport pathways (Z-scheme) in plant photosynthesis

lines of 5 ms/35 °K [33] or 2 ms [34] have been attributed to the photosystem II reaction centre $P680^{+\cdot}$. The ejected electrons are transferred to the electron carriers which form the electron transport chain (Fig. 3). $P680^{+\cdot}$ is reduced by water, via a water splitting mechanism thought to involve a manganese enzyme (designated Y or Z), or other suitable artificial electron donors. In recent years extensive studies have focussed on artificial electron transfer processes for the purpose of harnessing the generated photocurrent either for direct use or for conversion to secondary storage forms such as H_2, H_2O_2 or carbon reduction. Photovoltaic devices which incorporate chlorophyll layers or aggregates immobilized onto conducting surfaces have been shown to mimic the primary photochemical processes of P680 and P700 (Table 1), although at low efficiencies. Alternatively charge transfer across artificial biomembranes such as monolayers [35–44], micelles [45–56], liposomes or vesicles [57–66], microemulsions [67–68], has been demonstrated. These systems organize donors and acceptors, lower ionization potentials, and most importantly, through their interfaces or electrical double layers, allow for some kinetic control of electron transfers. In a recent study by Toyoshima et al. [65] with chlorophyll-a incorporated liposomes has shown that reduction of copper(II) in an aqueous phase could be achieved by illuminating vesicles containing potassium ascorbate. The reduction of copper(II) to copper(I) could be monitored by the loss of copper(II) e.p.r. activity.

In plants electron transfer is mediated through membrane bound electron carriers arranged in series to form what is termed the Z-scheme (Fig. 3). The characteristics of the Z-scheme have previously been reviewed in detail and only a brief outline of the chemical intermediates will be given here [69–77]. The essential feature of the model, from the point of view of this review, is the evolution of molecular oxygen as a result of redox reactions involving manganese since this part of the photosynthetic scheme is the least understood.

2.1 Cytochromes

A detailed description of the structure and function of cytochromes is given in recent reviews [78,79]. Three different cytochromes have been isolated and purified from higher plant chloroplasts. Two of the three purified plant chloroplasts cytochromes, b_6 and b-559, are defined to be b-type since the heme can be split from the protein component by acid-acetone treatment and the reduced α-band maximum of the pyridine hemochromagen is approximately 556 nm. The third cytochrome is c-type in terms of a thioether heme linkage to cysteines of the apoprotein through positions 2 and 4 of the porphyrin rings. The b-type cytochromes have typical molecular weights (M.W./heme) approaching 40,000 D (cyt-b_6) and 46,000 D (cyt-b559) although they are generally purified as multisubunit lipoproteins of molecular weight 60,000 D and 111,000 D respectively. Of the lipid content of cyt-b_6, 75–80% of the lipid is accounted for by seven chlorophyll-a and six cardiolipin molecules per heme. The 40,000 D protein component consists of separable subunits of 20,000, 96,000 and 6,600 (× 2) D. The total molecular weight of cyt-f is uncertain due to aggregation phenomena although photoactivating species of molecular weight 35,000 D and 65,000 D have been prepared.

Table 1. Photovoltaic devices involving chlorophyll

Species	Electrode	Comments	Ref.
(Chl.a. $2\,H_2O)_n$	Pt | Chl.a H_2O | | H_2O | Pt	1. positive photopotential generated 2. O_2 and H_2 detected 3. half cell reactions a) Chl.a-photocathode $2\,H_2O + 2\,e^- \rightarrow H_2 + 2\,OH^-$ $E_0 = -0.42\,V$ b) Chla-free anode $2\,H_2O \rightarrow 4\,H^+ + O_2 + 4\,e^-$ $E_0 = -.81\,V$	1–6
Chl.a/b	Stearic acid monolayers of Chl.a.b deposited onto SnO_2 optically transparent electrodes (OTE)	1. anodic photocurrent developed 2. negative photovoltage developed 3. quantum efficiency 12–16%	7
Chl.a	Pt | liquid crystal | Chl | Buffer soln. | | Buffer soln. | Pt	1. cathodic photocurrent developed 2. positive photovoltage shift 3. decomposition of H_2O, H_2 gas detected	8
(Chl.a. $2\,H_2O)_n$	Chl.a | Pt	CO_2 assimilation $2\,CO_2 + 2\,H_2O \xrightarrow[\text{Chla}]{h\nu} 2\,H_2CO_2 + O_2$	9
Chl.a	Al | Chl.a | Ag	Efficiency a) 0.036% if anhydrous Chl.a b) 0.21% if crystalline Chl.a	10

Table 1 (continued)

Species	Electrode	Comments	Ref.
Chl.a	E(Pt∣Chl.∣NQ∣NAD^+∣∣ $Fe(CN)_6^{4-}$∣E(Pt∣Chl.∣AHQ) NQ – Naphthoquinone AHQ – Anthraquinone	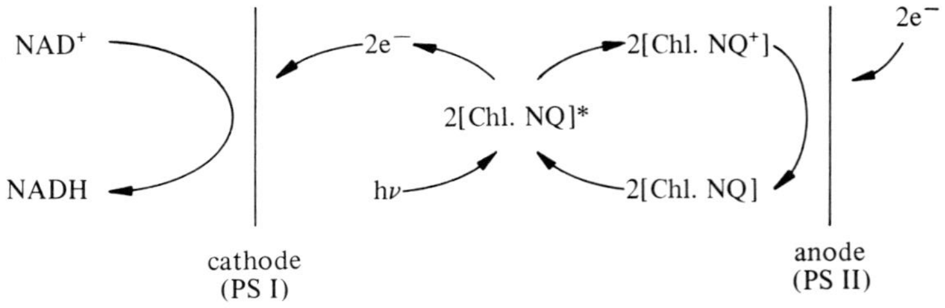 The photosensitizer, naphthoquinone (NQ), acts as an electron acceptor in the cathodic process.	11
	Chl.a∣Quinone∣Pt∣β-carotene TCQ – Tetrachloroquinone	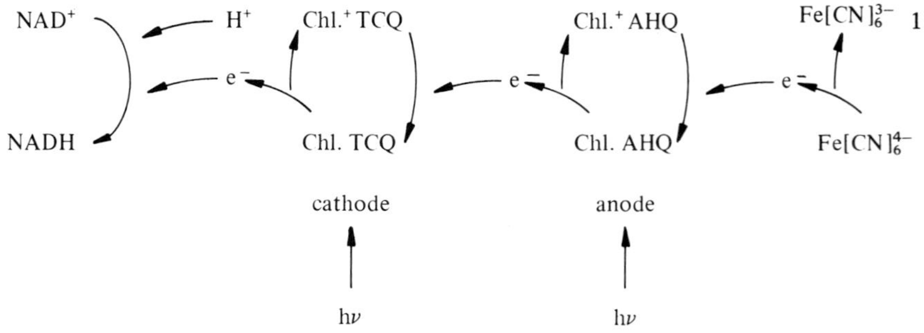	12
Chl.a	Pt, Chl.a∣Chl. a^+∣$LiClO_4$∥ NaOH, $NaHCO_3$∣H_2Q/Q, Pt	1. Positive photopotential generated 2. Charge transfer process analogous to p-type semiconducting films	13

Chl.a	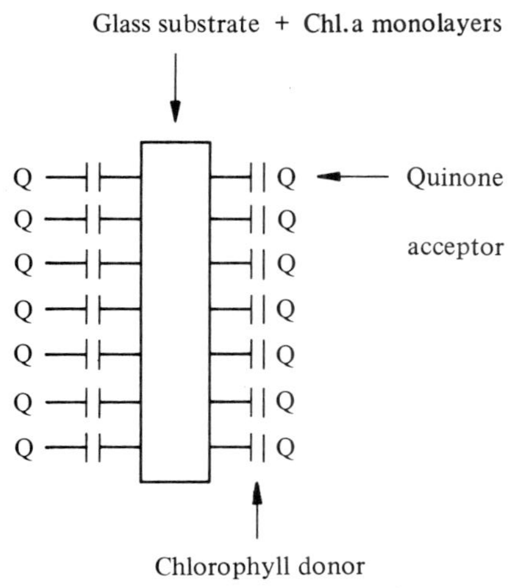	1. Under steady state illumination, $Chl.a^{\cdot +}$ with e.p.r. activity g = 2.0025, is generated. 2. Magnetic circular dichroism showed light induced bleaching at 680 nm.	14
	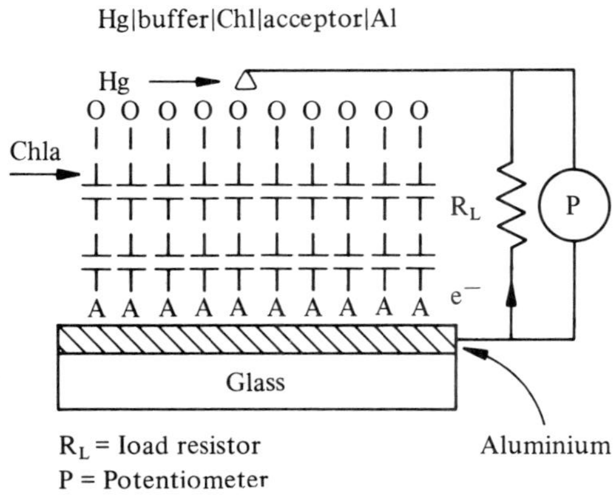	1. Without acceptor, open circuit voltage of 150 mV generated at illumination ~ 1000 Wm^{-2} with a measured photocurrent of 34 nA cm^{-2}. Energy conversion efficiency of 4×10^{-5} based on red photons absorbed. 2. With acceptors, plastoquinone/stearic acid (10%), open circuit voltage of 270 mV generated at illumination — 1000 Wm^{-2} with a measured photocurrent of 140 nA cm^{-2}. Energy conversion efficiency of 4×10^{-4} based on red photons absorbed.	15

Table 1 (continued)

Species	Electrode	Comments	Ref.
Chl.a/b	E(Chl-liquid crystal) | buffer || buffer | pt liquid crystal; COL: Cholesteryl oleate MBBA: N-(p-methoxybenzylidene)-p-butylaniline HCB: 4-Heptyl-4-cyanobiphenyl	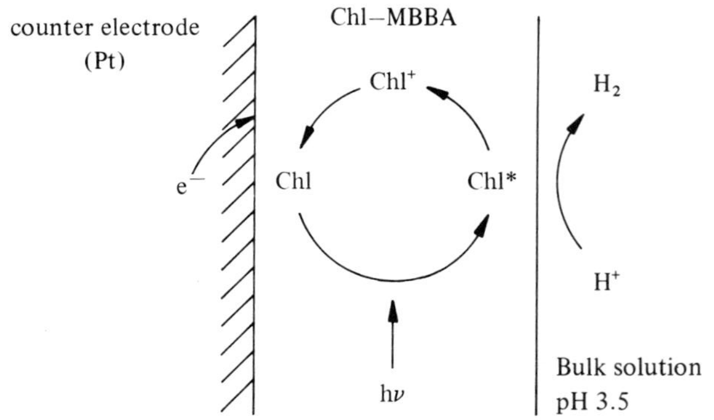 1. E(Chl-MBBA) generated a photovoltage of 80 mV and a cathodic photocurrent. 2. H_2 gas evolved from the bulk solution.	16

Mn.Chl.a	E(Mn.Chl-HCB) |buffer || buffer |pt	see diagram and notes below	17

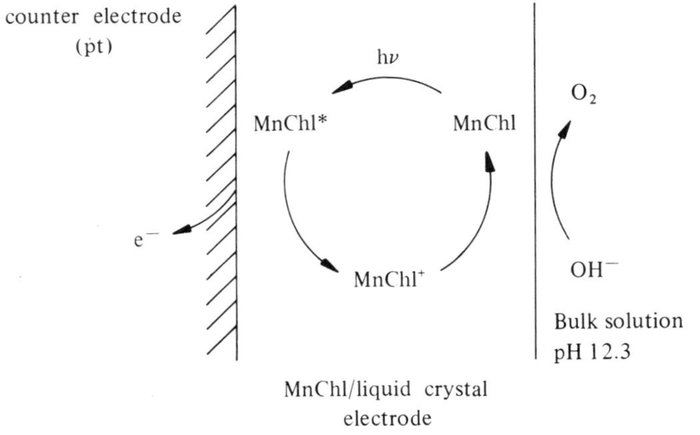

1. Molecular oxygen from water was generated.
2. Upon illumination a negative photovoltage shift suggesting a flow of electrons from water to the platinum electrode.
3. The role of the liquid crystal is to organize the light trapping chlorophyll and to facilitate charge transfer in a manner undertaken by the membrane in the chloroplast.

1. Fong, F. K., Winograd, N.: J. Am. Chem. Soc. *98*, 8, 2289 (1976)
2. Fong, F. K., et al.: ibid. *99*, 17, 5802 (1977)
3. Fong, F. K., Hoff, A. J., Brinkman, F. A.: ibid. *100*, 2, 619 (1978)
4. Fong, F. K., Galloway, L.: ibid. *100*, 11, 3594 (1978)
5. Galloway, L., et al.: ibid. *101*, 1, 229 (1979)
6. Galloway, L., et al.: ibid. *100*, 14, 4635 (1978)
7. Miyasaka, T., et al.: ibid. *100*, 21, 6657 (1978)
8. Aizawa, M., et al.: Biochim. Acta *23*, 1061 (1978)
9. Fruge, D. R., Fong, G. D., Fong, F. K.: J. Am. Chem. Soc. *101*, 13, 3694 (1979)
10. Dodelet, J. P., Le Brech, J., Le Blanc, R. M.: Photochem. Photobiol. *29*, 1135 (1979)
11. Takahashi, F., et al.: Electrochim. Acta *22*, 289 (1977)
12. Takahaski, F., Kikuchi, R.: Bull. Chem. Soc. Jap. *49*, 3394 (1976)
13. Fong, F. K., Winograd, N.: J. Am. Chem. Soc. *98*, 2287 (1976)
14. Janzen, A. F., Bolton, J. R., Stillman, M. J.: J. Am. Chem. Soc. *101*, 6337 (1979)
15. Janzen, A. F., Bolton, J. R.: ibid. *101*, 6342 (1979)
16. Aizawa, M., et al.: Electrochim. Acta *23*, 1061 (1978)
17. Aizawa, M., Hirano, M., Suzuki, S.: ibid. *24*, 89 (1979)

The function of the cytochromes in electron transport may be in the coupled synthesis of ATP. Cyt-b559 may also play a role in water splitting and in the process of intramolecular proton movement. Two forms of cyt-b559 (λ_{max} = 559 nm) have been detected, a) a high potential form cyt-b559$_{HP}$, midpoint potential at pH7 +0.383 V [78] or +0.370 V [80] and b) a low potential form cyt-b559$_{LP}$, midpoint potential at pH7 +0.077 V [78] or 0.020 V [80]. Conversion of the higher potential to a lower potential form occurs via treatment with membrane disturbing agents such as detergents. There is disagreement as to whether their high and low potential forms are separate protein components or different potential states of the same protein with the heme in different ionic or polar environments. According to a scheme proposed by Cramer [78], cyt-b559$_{HP}$ may function in a cycle around PS II contributing to water splitting. A reversible light induced negatively directed potential change gives rise to cyt-b559$_{LP}$ whose functional site is located between the primary electron acceptor of PS II, and the plastoquinone pool. Treatment of chloroplasts with hydroxylamine gave rise to two new forms of cyt-b556 titratable with hydroquinone (i) $Em_{7.8}$ = + 240 mV and (ii) $Em_{7.8}$ = + 90 mV. Upon allumination only the lower potential species could be detected. The loss of higher potential cyt-b559 as a result of NH_2OH treatment correlated perfectly with the effects on oxygen evolution from water photo-oxidation [81], implicating a possible role and functional site of cyt-b559 in photosystem II reactions [82]. Recently Hoffmann et al. [83] have suggested that cyt-b559 may function as an electron transfer intermediate in a pathway by passing the inhibitor 3-(3,4-Dichlorophenyl)-1,1-dimethylurea (DCMU), which is believed to block electron transfer between the primary electron acceptor and the secondary electron acceptor (plastoquinone pool) of photosystem II, [84] and regulating the direct transfer of electrons from the acceptor side of PS II to an artificial electron acceptor Dichloroindophenol (DCIP).

Cytochrome b_6 (midpoint potential – 0.08 V) appears to be involved in a photosystem I cyclic phosphorylation, ATP forming, cycle with plastoquinone having the likely electron acceptor [78]. Cytochrome-f (midpoint potential + 0.365 V) is thought to be closely linked to the coupling mechanism of ATP formation.

2.2 *Primary Electron Acceptor of Photosystem II*

The chemical identity of the primary electron acceptor of photosystem-II has not yet been established although extensive spectroscopic evidence has suggested three possible candidates [77,85]. The proposed electron acceptors are; (i) a component, designated Q (Quencher), measured by its quenching of photosystem II fluorescence and having a midpoint potential of – 35 mV; (ii) a component, designated X-320, measured by a light induced absorbance change at 320 nm which was identified with the univalent reduction of plastoquinone to the semiquinone anion, although attempts to detect the radical form by e.p.r. methods have proved fruitless; and (iii) a component, designated C550, measured by a light induced absorbance change near 550 nm which is thought to be secondary and may serve as an indicator of the redox states of the primary electron acceptor. The absorption change kinetics of the one electron reduction of X320 (a special plastoquinone) to a semiquinone radical anion parallel the kinetics of the fluorescence quencher Q. The identification of X320 with Q is supported by electrochemical similarities.

Although the majority of opinion suggests Q (X320) to be the primary electron acceptor of photosystem II, recent investigations have indicated that additional electron acceptors mediate the transfer charge from P680 to plastoquinone (Q). The free base adduct of chlorophyll, pheophytin has been identified by optical, redox and paramagnetic resonance measurements as a likely primary photo reduced species[85a,86,86a).

$$\text{P680} + \text{Pheophytin} \rightarrow \text{P680}^{+\cdot} + \text{Pheophytin}^{-\cdot} .$$

With an estimated reduction potential of -640 mV, Pheophytin$^{-\cdot}$ is easily capable of reducing Q (X320).

2.3 *Plastoquinone*

Plastoquinone consists of several related substances of which the most abundant is plastoquinone A. Plastoquinone A is a substituted benzoquinone found almost exclusively in algae and higher plants in relatively high concentration (spinach 100 μmol mmol^{-1}).

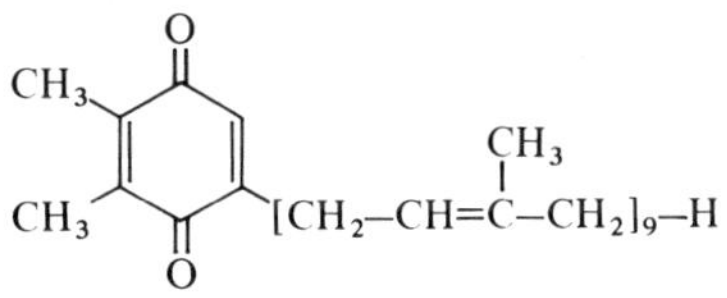

Plastoquinone (λ_m = 260 nm) undergoes reduction to the hydroquinone (λ_m = 290 nm (w)) via a two electron process corresponding to a midpoint potential of + 113 mV. One electron reduction leads to a semiquinone free radical which shows a characteristic strong absorbance of about 320 nm. Spectral investigations of endogenous plastoquinone show that upon illumination plastoquinone is reduced to plastohydroquinone (E_m = + 80 mV) whilst cytochrome-f remains mainly in the oxidized form[87).

2.4 *Plastocyanin*

Plastocyanins are relatively low-molecular weight "blue" copper proteins[88) found in photosynthetic organisms[89). It was originally proposed that spinach plastocyanin is a dimer of molecular weight 21,000 Daltons containing two copper atoms per monomer. Plastocyanin found in *Phaseolus vulgaris* is a monomer of molecular weight 10,690 Daltons containing a single copper atom. The amino acid sequences of plastocyanins show the protein to contain one cysteine and two histidines in homologous and highly conservative regions[90]. The coordination site of the copper ion has recently been identified by X-ray crystallography[90a]. The copper is located in a distorted tetrahedral environment which is coordinated by the thiol group of Cys-84, the thioether group of Met 92 and the imidazole groups of His 37 and His 38. Membrane bound plastocyanin has a midpoint potential of + 320 mV (soluble protein E_m = + 370 mV) and is thought to be the electron donor from cytochrome-f. (Rate of reduction, $k = 3.6 \times 10^7\ M^{-1}\ s^{-1}$) to $P700^+$ (Rate of oxidation, $k = 8.0 \times 10^7\ M^{-1}\ s^{-1}$)[89].

The e.p.r. of oxidized plastocyanin (Cu^{2+}) show resonances $g_\perp = 2.05$ and $g_{\parallel} = 2.23$ with $A_{\parallel} = 0.006\ cm^{-1}$[88]. Reduction of plastocyanin gives rise to an e.p.r. silent species (Cu^+). Recent NMR studies of plastocyanins from French bean *(Phaseolus vulgaris)* and cucumber *(Cucumis sativus)* have identified two binding sites towards artificial redox indicators, $Cr(CN)_6^{3-}$ and tris(1,10-phenanthroline)-Cr^{III}. The binding of the complexes to plastocyanin occur at sites separated by 15 Å with one site lying in close proximity to the copper atom[91,92]. This observation confirms recent X-ray and rate constant studies which suggested that two binding sites are involved in electron transfer reactions to and from the copper atom and that possibly two mechanisms of electron transfer are involved[90a,93].

2.5 *The "Rieske" Iron-Sulphur Centre*

Electron transfer from reduced plastoquinone to oxidized cytochrome-f is thought to occur via a membrane bound Fe–S protein referred to as the "Rieske" iron-sulphur centre after the discovery of an Fe–S protein in mitochondrial and submitochondrial fragments by Rieske et al.[94,95] which exhibits an e.p.r. signal (g = 1.89) similar to that later found by Malkin et al. in chloroplasts[96]. The e.p.r., redox properties and function of Rieske's protein have been recently reviewed[97,98]. The juxtapositioning of the Rieske Fe–S centre (midpoint potential + 0.290 V) with the plastoquinone pool is supported by recent studies which show inhibition of electron transfer to the Fe–S centre by Dibromothymoquinone (DBMIB E_m + 0.170 V), a known plastoquinone protagonist[99–102]. The addition of DBMIB to chloroplasts treated with an electron donor, Duroquinol, results in a loss of e.p.r. activity at g = 1.89 with the appearance of an unidentified species at g = 1.94. Another inhibitor, 3-(3,4-dichlorophenyl)-1,1-dimethylurea (DCMU), which functions between the primary electron acceptor of PS II and plastoquinone has no effect on the reduction of the Fe–S centre with Duroquinol. On the basis of these results together with a consideration of the midpoint potential of cytochrome-f (+ 360 mV) the following scheme has been proposed[103–105].

Recent e.p.r. studies show a preferred orientation of "Rieskes" Fe–S protein from pigeon heart mitochondria with $g_y \approx 1.90$ in the plane of the orientated membranes and $g_x \approx 1.81$ at 90°. The z-axis of the g-tensor runs through the two Fe atoms and probably lies in the plane of the membrane[106]. Similar experiments have been performed on the spinach chloroplast Fe–S centre showing that the protein has axial symmetry with $g_{\parallel}$ close to the plane of the membrane[107].

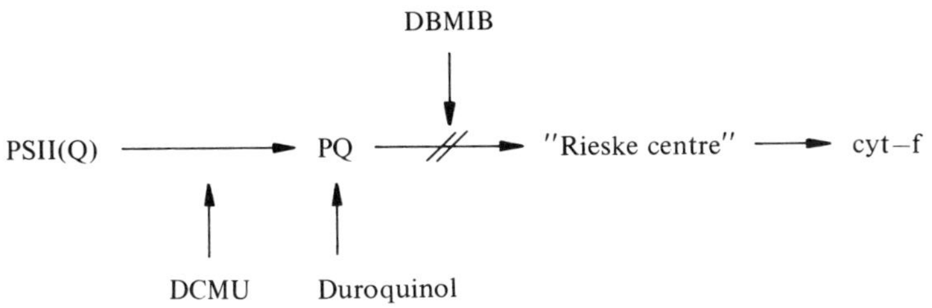

2.6 *Primary Electron Acceptor of Photosystem I*

The properties that the primary electron acceptor of photosystem I should exhibit have been outlined by Golbeck et al.[75]. These include (i) the light-induced reduction kinetics should mirror the oxidation kinetics of the primary donor, P-700, (ii) the kinetics of the back reaction between the reduced acceptor and oxidized donor should be identical for both species; and (iii) a kinetic correlation should exist between the reoxidation of the primary acceptor and reduction of known secondary acceptors. Several species have been proposed which include chlorophyll, pteridines, flavins, ferredoxin and porphyrins (Ref. 75, Table 1). Evidence for an intermediate electron carrier preceding P-430 has come from recent studies of flash-induced transient absorption changes[108,109]. The intermediate, designated A_1, accepts an electron from reduced P700 which is then transferred to another intermediate A_2 (or X). The chemical identity of A_2 is thought to be a chlorophyll-a. Evidence supporting this assignment comes from comparison of the PS I optical different spectra with those of *in vitro* generated $Chl^{\overline{\cdot}}$ radical anions in solution[86]. The *in vitro* potential of $Chl^{\overline{\cdot}}$ (midpoint potential –0.900 V) is sufficient to reduce the bound ferredoxins, (midpoint potential – 0.730 V).

Spectral and kinetic evidence recently reported suggests A_1 may be a chlorophyll dimer which is located between P700 and A_2 (X) and whose function is to transfer an electron from $P700^+$ to A_2 (X)[110]. More recently, proposals suggest monomeric chlorophyll or possibly pheophytin as the likely intermediate[111].

2.7 *Chloroplast Iron-Sulphur Centres*

Iron-sulphur centres and their associated protein component are ubiquitous in living organisms, and they exhibit a variety of functions that are always related to electron transfer processes.

Recently the biological activity of synthetic iron-sulphur, $[Fe_4S_4(S \cdot CH_2CH_2\text{-}OH)_4]^{2-}$, and iron-selenium, $[Fe_4S_4(S \cdot CH_2CH_2OH)_4]^{2-}$, structural analogues of the ferredoxin $[Fe_4S_4]$ active centre has been investigated[112,113]. The clusters could replace ferredoxin and mediate electron transfer to *Clostridium pasteuridium* hydrogenase in a H_2-evolving system with sodium dithionite as the electron donor.

The structure, function, and spectroscopic properties of the naturally occurring Fe–S centres have attracted a large number of investigations[97,114–118a]. Among the iron-sulphur proteins two distinct classes of these proteins transfer electrons at oxidation-reduction potentials near the potential of the H_2/H^+ couple. These are: (1) the ferredoxins of plant origin that contain a cluster of two non-heme and two acid labile sulphur atoms per molecule $[Fe_2S_2(cys)_4]^{2-}$ and (2) the bacterial type ferredoxins which contain one or two clusters of four iron, four labile sulphur atoms per molecule $[Fe_4S_4(cys)_4]^{2-}$. Recent e.p.r., Mössbauer, and X-ray diffraction studies of ferredoxins from *Desulforibrio gigas* [119] and *Azobacter vinelandii* [120–122] suggest a new novel metallo centre containing a ring of three iron atoms linked by three labile sulphur atoms.

Redox properties of Fe–S centres have been extensively studied by e.p.r. at temperatures below ~ 80 K [75,97,118]. These signals are caused by molecular antiferromagnetism between a high spin iron(III) and a high spin iron(II) which gives rise to an S = 1/2 system exhibiting a characteristic principle e.p.r. g-value of 1.94. Reduced iron-sulphur centres in photosynthetic systems display either rhombic ($g_x \neq g_y \neq g_z$) or axial ($g_\perp \neq g_\parallel$) e.p.r. spectra. In conjunction with e.p.r., redox potentiometry affords a method of distinguishing multiple iron-sulphur centres on the basis of their midpoint redox potentials which may serve different roles in electron transport processes.

It is thought that two types of Fe–S centres may be involved in chloroplast photosystem I processes (a) Ferredoxin, which is involved in the non-cyclic electron transfer pathway to NADP, and (b) at least two Fe–S proteins (centres A and B) which are involved in electron transfer reactions prior to ferredoxin. The two types of Fe–S centres are distinguishable on the basis of solubility; ferredoxin being the soluble centre whilst the remaining centres are insoluble or bound. The two well characterized iron-sulphur centres A and B are of low oxidation-reduction mid-point potentials of – 550 mV and – 590 mV respectively [123,124] and comprise the species, P-430, which was previously assigned as the primary electron acceptor of P700 [71]. The photoreduced centres A and B, from spinach, exhibit different e.p.r. signals at 77 °K with g values of 2.05, 1.94, 1.86 and 2.07, 1.92 and 1.89 respectively. Microwave power saturation and temperature dependent studies of the e.p.r. signals of dithionite or ascorbate reduced centres suggest a magnetic interaction occurs between centres A and B [125]. The magnetic interaction between the [4 Fe–4 S] centres of the electron acceptors A and B resulted in saturation properties similar to those of the 2 [4 Fe–4 S] ferredoxin from *Clostridium pasteurianum*. The assignment of the e.p.r. signals of centres A and B to paramagnetic 4 Fe–4 S centres similar to those found in reduced ferredoxins, on the basis of the similarity of the spin states of the iron-containing centres in the two cases, has been confirmed by Mössbauer studies of photosystem I preparations from the blue-green algae *Chlorogloea fritschii* [126].

Two tentative proposals have been suggested for the role of centres A and B [97].

(i) $P700 \rightarrow A_1 \rightarrow$ "X" $\rightarrow$ Centre B $\rightarrow$ Centre A $\rightarrow$ Ferredoxin

and

(ii) $P700 \rightarrow A_1 \rightarrow$ "X" $\nearrow$ Centre B $\rightarrow$ cyclic electron transport
$\searrow$ Centre A $\rightarrow$ non-cyclic electron transport

The chemical identity of the electron acceptor, X (midpoint potential −0.735 V), is currently unknown although it is thought that X is identical to an electron acceptor designated A_2 (midpoint potential −0.730 V) which shows a difference spectrum resembling that of a bound Fe–S protein [110]. A similar suggestion has been put forward by Cammack et al. [127] who identified the reduced acceptor, X, by e.p.r. spectroscopy, g = 1.78, 1.88 and 2.08 of PS I centres in intact cells of cyanobacteria (blue-green algae) *Nostoc muscorum* and *Anabaena cylindrica.*

An alternative proposal had just recently emerged from a study of the electron spin relaxation of centre X from spinach photosystem I particles by microwave power saturation of e.p.r. signals which showed that centre X does not exhibit spectral features of known typical [2 Fe–2 S] or (4 Fe–4 S) centres [125]. It was concluded that centre X is most probably an anion radical of chlorophyll which interacts with a fast relaxing ion, most probably iron, which is not associated with either centres A or B.

3 Kinetics of Oxygen Evolution

The formation of a molecule of oxygen from water requires the removal of four electrons; the average redox potential of these four equivalents is about +.81 V. However, the four individual reaction steps in the chemical oxidation of water vary from −0.45 to +2.33 V. Since the energy of a red photon is about 1.8 eV, some of these oxidation steps might be energetically unfavourable without a "moderating" mechanism. Some aspects of oxygen evolution kinetics have been discussed in previous reviews [128–133].

The development of the modulated polarographic technique by Joliot has permitted a refined analysis of the priming of oxygen evolution by light [134]. The technique was successfully applied by Joliot et al. [135] and Kok et al. [136] and a detailed analysis of the yields of oxygen per short light flash were made.

With appropriate spacing (~ 300 ms) between saturating light flashes (2 μs half amplitude) these workers observed that the yield of oxygen per flash was maximal with the third flash and with subsequent flashes, large oscillations of yields were obtained and these maxima occurred with a period of four (Fig. 4). A number of kinetic models have been proposed to explain this observation [135–137]. As outlined by Diner and Joliot [133], the model which best explains the existing data is that proposed by Kok et al. [136,138] in which oxidizing equivalents are successfully accumulated at an oxygen evolving site attached to only one photosystem II reaction centre. Oxidizing equivalents are neither exchanged between different photosystem II centres nor between oxygen producing sites. Each single turnover flash generates one oxidizing equivalent and the accumulation of four of these results in the liberation of an oxygen molecule. The cycle then begins again.

The characteristics of the model are depicted in Fig. 5 which shows that driven by four successive photoacts, the oxygen-evolving enzyme (Y) cycles through five oxidation states designated S_0, S_1, S_2, S_3 and S_4. The temperature dependancc of

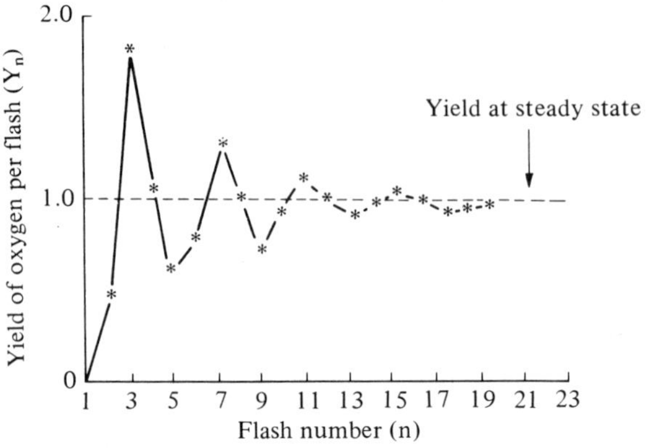

Fig. 4. Pattern of oxygen evolution resulting from short saturating flashes

the photo-oxidation steps, $S_n \rightarrow S_{n+1}$, has recently been studied by comparing thermoluminescence oscillation patterns of spinach chloroplasts as a function of the number of short saturating light flashes[139]. It was found that the S_0 or $S_1 \rightarrow S_2$ reaction proceeds below −65 °C whereas $S_2 \rightarrow S_3$ and $S_3 \rightarrow S_4$ transitions are blocked at temperatures below −35 °C and −20 °C respectively.

Each single turnover flash induces a separation of charge, transforming centres in redox state $Chl(Q)S_n$ to redox state $Chl^+(Q)^-S_n^*$, where Q represents the primary

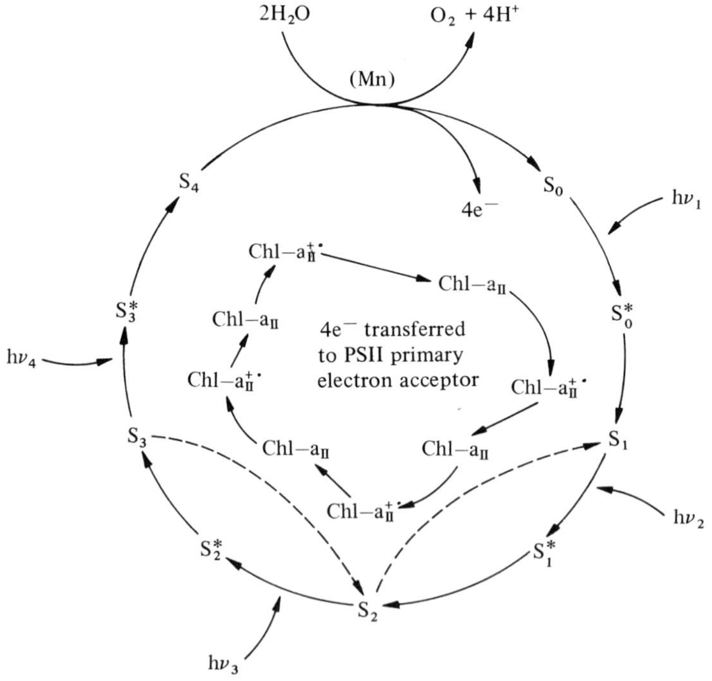

Fig. 5. S-state mechanism of oxygen evolution

electron acceptor of $P680^+$ (or $Chl.a_{II}^{+\cdot}$)[133]. The positive charges generated are stored by the dark relaxation $S_n^* \rightarrow S_{n+1}$ which corresponds to the oxidation of S_n by $Chl.a_{II}^{+\cdot}$ via univalent electron transfer reactions until after the accumulation of four redox equivalents, S_4 decomposes to S_0, and water cleavage by electron abstraction occurs and oxygen is evolved[140].

The oscillatory pattern of oxygen flash yield is damped and eventually attains a non-oscillatory steady state. To account for the pattern behaviour two damping phenomena were postulated; (1) "misses", α, in which there are randomly recurring failures of the photochemical charge separation, and (2) "double hits", β, which is defined as the occurrence of two charge separations, $S_n^* \rightarrow S_{n+2}$, in a single flash[138,141]. It is thought that α (probability factor) reflects a biological rather than technical phenomenon[131,132]. In isolated chloroplasts α is usually 0.10–0.12, while in whole cells (e.g. Chlorella) $\alpha = 0.20$–0.25. Double hits, β, probably reflects a technical limitation in the flash yield measurements and is of the order ~ 0.05[131].

To be consistent with the experimental observations the model carries the additional assumption that S_1 and S_0 are both stable in the dark. The lifetimes of the S_2 and S_3 states are long compared to the dark transitions ($S_n^* \rightarrow S_{n+1}$) but less than about 15 min: they decay to S_1. Consequently after about a 15 min dark period the distribution of the S-states is $S_0:S_1:S_2:S_3 = 0.25:0.75:0:0$. The ratio S_1/S_0 can be varied by a number of ways including preillumination and by chemical means such as ferricyanide (S_1/S_0 becomes very high) and ascorbate-DCPIP which reduces S_1/S_0 towards unity.

In recent years a detailed analysis of Kok's model for oxygen evolution has been reported by Lavorel[142,143], Thibault[144] and Jursinic[145]. Lavorel has developed a numerical procedure (σ-analysis) for the objective evaluation of the damping coefficients of oxygen sequences. The authors' results argue against the concept that misses (α) are only of photochemical origin, as proposed in Kok's model. The application of σ-analysis (the application of matrix analysis to the model allows one to derive three symmetrical functions (σ_1, σ_2, σ_3) of the transition probabilities of the model) by Lavorel led to an alternative to Kok's model for the oxygen-evolving system in photosynthesis. The model illustrates a more rigorous treatment of the damping phenomena which have been observed in the measurement of oxygen flash yield (Y) whilst maintaining the concept of a four step mechanism for oxygen production. Thibault has carried out a critical examination of the σ-analysis method and showed that the model was extremely sensitive to experimental error. Furthermore, the matrix on which the σ_i's calculation is based is an ill-conditioned matrix, a property of which is to amplify errors. Consequently, the author cautions that conclusions drawn from the procedure must be accepted with care[144].

A mathematical model, based on the binomial distribution, for predicting activation of reaction centres at various flash intensities from supersaturating flashes to intensities two orders of magnitude lower, has recently been proposed[145]. Evidence is presented suggesting that the photosystem II photosynthetic unit is quite inhomogeneous, that is having a variety of antennae sizes and that α and β are nearly independant of flash intensity and are intrinsic properties of the oxygen-evolving system. Although their mechanistic origin is unknown, α and β are probably related to special properties of charge flow in photosystem II.

3.1 Proton Release During Oxygen Evolution

Concomitant to the release of molecular oxygen, protons arising from water are released in each of the S-state transitions, except $S_1 \rightarrow S_2$. A similar oscillation pattern has been reported for the release of protons by a series of flashes which point to a concerted reaction in which protons are released in synchrony with oxygen [146]. The protons generated do not diffuse freely as does oxygen but rather are readily taken up by ionizable groups, exchange with counterions, and do not readily diffuse through membranes. Kok and Fowler [146] used a sensitive pH technique to measure protons released by short light flashes. They found that proton release, coupled to the production of oxygen and thus released from the oxygen-evolving site, occurs toward the inside of the thylakoid membrane. The result supports the location of the photosystem II donor side on the inner surface of the thylakoid membrane.

Even though a considerable amount of kinetic information is available on the oxygen-evolving mechanism, our knowledge of the detailed chemistry of the S-states model is very inadequate. We do know however that there exists an essential requirement by the chloroplast for inorganic ions such as manganese(II) and chloride for oxygen production. Investigations of this requirement has led to the proposal that the oxidation of the S-states of the water splitting enzyme is related to redox reactions of manganese *in vivo* involving Mn(II), Mn(III) and possibly Mn(IV).

3.2 Abundance and Properties of Chloroplast Manganese

A starting point to the supposition that manganese has a unique function in photosynthesis arose from an early observation that manganese deficiency in a species of green algae, *Ankistrodesmus*, decreased photosynthesis without having significant effects on respiration or total chlorophyll [147]. Earlier studies had shown that manganese was an essential micro-nutrient growth factor for Chlorella species and it was thought that manganese was associated with the Hill reaction (O_2 evolution) [148–152]. Kessler et al. [153–155] and Eyster et al. [156] studying oxygen evolution inhibition under manganese deficiency had shared a similar opinion. Cheniae and Martin showed that the removal of Mn (< 1/1000 Chl) from normal Mn-containing chloroplasts by mild temperature shock resulted in a loss of oxygen-evolving capability, but did not affect System I reactions [132,157]. The early literature covering manganese involvement in oxygen evolution has been previously reviewed [128,158,159].

An earlier assessment of the abundance of manganese in chloroplasts revealed a wide spread of data [159]. The current accepted value is 1 g atom of bound manganese per 50–100 Chlorophyll molecules for the pool of manganese in PS II [116]. This pool of bound manganese is the minimum amount of Mn within PS II that is required for maximum quantum efficiency and high rates of O_2 evolution. Factors including sample preparation and plant age contribute to the variability of the manganese content. Kaniuga et al. have recently determined the manganese content of expanded tomato leaves, by complexiometric formation with pyridyl-2-azonaphthol and by the permanganate colorimetric method, to be 3 g-atom/400 Chl. Whilst young tomato leaves were found to have a ~ three-fold amount of manganese, 6–8 g-atom/ 400 Chl (Table 2, Ref. 2).

Another factor which may contribute to the accurate determination of manganese content is the heterogeneous binding nature of chloroplast manganese and the effectiveness of removing bound manganese by a number of chemical and physical treatments as outlined in Table 2. Tris washing of chloroplasts has been used extensively to interrupt the reactions linking the S-states of oxygen evolution to the system II reaction centre. Tris washing of chloroplasts removes about two-thirds of the total manganese pool whilst more drastic treatments such as temperature shock can remove all of the bound manganese.

Blankenship and Sauer[160,161)] showed that chloroplast bound manganese released by Tris (2/3 Mn pool) is released towards the inner thylakoid matrix and after some time ($t_{1/2} \approx 2.5$ h) diffuses through the membrane. E.p.r. investigations showed typical spectra of free manganese which diminished upon incubation with DCIP and ascorbate in sucrose buffers in the dark. Exogenous manganese, however, could only be re-inserted when the manganese depleted chloroplasts were illuminated. Chelation of loosely-bound manganese by EDTA ($K_d \approx 10^{-14}$) within the thylakoid did not inhibit re-insertion whereas the chelation of exogenous manganese inhibited the light driven uptake of manganese. It was concluded that manganese is taken across the membrane by a light driven process and that once within the thylakoid, is readily bound ($K_d \approx 1.2 \times 10^{-4}$ M) and that this binding is not dependant on light. The results support previous studies which similarly showed that manganese incorporation into manganese deficient algae proceeds via a multiquantum process[162,163)].

Kaniuga et al. found that upon cold and dark storage to tomato leaves between 40 and 50% of the total manganese is lost and following illumination of such leaves

Table 2. Removal of manganese from chloroplasts by various treatments

Treatment	Effect	Ref.
Thermal shock:		
1. Incubation at 45–50 °C for 3–10 min.	1) Complete removal of bound manganese. 2) Total reduction of Hill reaction activity.	1
2. Incubation at 0 °C for up to 3 days in the dark.	1) Between 40–50% of the total manganese content is removed. 2) Hill reaction activity is reduced to 10–20% of untreated chloroplasts.	2–4
Hydroxylamine, NH_2OH Incubation at 15 °C with $[NH_2OH]$ = 2–10mM in the dark.	1) Between 65–75% of the total manganese content is removed. 2) Between 60–70% of the oxygen evolving capacity is inhibited. 3) Unprotonated hydroxylamine is the effective destructive agent. 4) Substituted hydroxylamine shows varying degrees of effectiveness according to: $NH_2OH \geqslant NH_2OSO_3 > CH_3NHOH >> NH_2OCH_3$	5

Table 2 (continued)

Treatment	Effect	Ref.
Sodium cyanide, NaCN Incubation at 5 °C with [NaCN] = 55mM, pH 8.0, for 10–20 h in the dark.	1) More than 80% of the bound manganese and copper, but not iron, is removed. 2) Hill reaction activity is lost.	6
Chaotropic agents[a] Incubation with either; $[NaClO_4]$ = 0.5M [NaSCN] = 0.8M [Guanidine] = 0.8M $[NaNO_3]$ = 1.0M [Urea] = 1.0M	1) Chaotropic agents inhibit electron transfer between water and photosystem-II. 2) Results suggest that membrane disorganization, giving rise to free Manganese(II) ions, may be involved.	7
Tris-HCl buffer[b] Incubation with 0.8M Tris-HCl, pH 8.0, for about 15 min.	1) Variable loss of total manganese reported ~ 70%[8,10], 30%[9,13] and 10%[11,12]. 2) Oxygen evolution capacity diminished upon manganese extraction.	8–13
Magnesium(II) ions, Mg^{2+} Incubation with $[MgSO_4]$ = 50–200mM, pH 7.9, at 0–2 °C for 30–90 min.	1) Up to 67% or two thirds of the total bound manganese content is removed. 2) The efficiency of oxygen evolution capacity is directly proportional to the bound manganese content.	14

a Chaotropic agent: A substance that enhances the partitioning of non-polar molecules from a non-aqueous to an aqueous phase as a result of the disruptive effect that the substance has on the structure of water. Chaotropic agents are generally ions, such as SCN^- and ClO_4^-, that have a large radius, a negative charge, and a low charge density; they are used to solubilize membrane bound proteins, to alter the secondary and tertiary structure of proteins and nucleic acids, and to increase the solubility of small molecules. (Dictionary of Biochemistry (Ed. J. Stenesh) (1975), J. Wiley & Sons.)

b Tris: Tris(hydroxymethyl)aminomethane

References for Table 2

1. Kimimura, M., Katoh, S.: Plant Cell Physiol. *13*, 287 (1972)
2. Kaniuga, Z., Zabek, J., Sochanowicz, B.: Planta *144*, 49 (1978)
3. Kaniuga, Z., Zabek, J., Michalski, W.P.: ibid. *145*, 145 (1979)
4. Michalski, W.P., Kaniuga, Z.: Biochem. Biophys. Acta *589*, 84 (1979)
5. Cheniae, G.M., Martin, I.F.: Plant Physiol. *47*, 568 (1971)
6. Takahashi, M., Asada, K.: Eur. J. Biochem. *64*, 445 (1976)
7. Lozier, R., Baginsky, M., Butler, W.L.: Photochem. Photobiol. *14*, 323 (1971)
8. Homann, P.: Biochem. Biophys. Res. Commun. *33*, 229 (1968)
9. Itoh, M., et al.: Biochim. Biophys. Acta *180*, 509 (1969)
10. Cheniae, G.M., Martin, I.F.: ibid. *197*, 219 (1970)
11. Yamashita, T., Tsuji, J., Tomita, G.: Plant Cell Physiol. *12*, 117 (1971)
12. Yamashita, T., et al.: ibid. *13*, 353 (1972)
13. Selman, B.R., Bannister, T.T., Dilley, R.A.: Biochim. Biophys. Acta *292*, 566 (1973)
14. Chen, K.Y., Wang, J.H.: Bio-inorg. Chem. *3*, 367 (1974)

the total manganese content in chloroplasts increases up to 70–80% of the manganese level measured in fresh leaves (Table 2, Ref. 2).

The reincorporation of manganese to chloroplasts treated with NaCN and the binding state of Mn has been studied by Takahashi et al. (Table 2, Ref. 6). It was found that $^{54}Mn^{2+}$ could be incorporated into two binding sites; (a) a low affinity binding site, $K_d = 10^4 \text{M}^{-1}$, incorporating 78% of added $^{54}Mn^{2+}$ and (b) a higher affinity binding site, $K_d = 2 \times 10^5 \text{M}^{-1}$ incorporating the remainder 22% of the $^{54}Mn^{2+}$. The high affinity binding of $^{54}Mn^{2+}$ was specific for NaCN-treated chloroplasts and the binding of $^{54}Mn^{2+}$ is inhibited by weak light though the low affinity binding site is not affected.

In the dark, several reductants induced the decrease of the Mn^{2+} binding by a mechanism similar to photo-inhibition[164]. A pH dependant structural change in the chloroplast membrane induced by light or redox agents is thought to disturb manganese binding by an apo-protein. The profile of pH dependance on the binding of manganese(II) was found to have a midpoint of pH 7.2 suggesting that a group with pk_a at 7.2 plays a role as the donor ligand of Mn^{2+} in the chloroplast membrane (Table 2, Ref. 6). An imidazole nitrogen of histidine and the α-amino groups of peptides has been proposed. Co-ordination of manganese(II) by thiol groups is excluded because the rebinding of manganese(II) was insensitive to a thiol reagent, *p*-Chloromercuribenzoate.

The binding characteristics of exogenous manganese by intact chloroplasts and chloroplasts treated with 0.8M Tris-HCl indicates a co-operative binding mechanism[31]. The nature of the co-operativity is not yet clear, although two hypotheses have been proposed: (1) The binding of the first manganese ions causes a change in the membrane structure that is favourable for further binding, and: (2) Binding is accompanied by change in the oxidation state of manganese, and stable oxidation and binding take place only when there are no less than two manganese ions at the given membrane locus.

3.3 Manganese Protein Complexes

In recent years interest has focussed on the binding of manganese by proteins and in particular the isolation of a manganese protein or enzyme thought to be involved in photosynthetic oxygen evolution. A number of manganese proteins from a variety of sources have been isolated and characterized. Some features of manganese proteins are outlined in Table 3. Of the Lectins, concanavalin A, a saccharide binding protein from Jack bean, has been extensively studied and shown to contain two binding sites: (1) a transition metal binding site, designated S_1, and: (2) a calcium ion binding site S_2. X-ray crystallographic studies have provided information pertaining to the structure around the co-ordination sites of manganese(II) and calcium(II). Manganese(II) occupies a very nearly octahedral co-ordination sphere by attachment to carboxylate groups of residues ASP 19, GLU 8, HIS 24 and the imidazolate function group of HIS 24 [165]. Manganese superoxide dismutases which function to protect cellular metabolic processes[166] have been detected in a number of bacterium and higher organisms. The structural comparisons of manganese superoxide dismutases with

Table 3. Manganese proteins

Protein and source	Function	Molecular weight (Daltons)	Manganese content	Properties	Ref.
Lectins: Concanavlin A Jackbean	Saccharide binding	26 000	1 atom of Mn/ protomer	1. Exists as a dimer at pH 6 and a tetramer at pH ⩾ 7 2. Protomer has two binding sites a) a site S_1 for transition metal ions (Mn^{2+}) and b) a site S_2 specific for calcium ions. 3. At low pH the metal ions dissociate with loss of saccharide binding capability	1–7
LBL_4 Lima beans		120 000	2 atoms of Mn^{2+}/ molecule of enzyme	1. Protein exists as a tetrameric unit 2. Tetramer can bind 2 Mn^{2+} or 4 Ca^{2+} 3. Ca^{2+} binding is co-operative 4. $LBL_4.Mn^{2+}$ can also bind 2 Ca^{2+} 5. $LBL_4.Ca^{2+}$ cannot bind Mn^{2+}	8
LBL_8		240 000	4 atoms of Mn^{2+}/ molecule of enzyme	1. Protein exists as an octameric unit 2. Octamer can bind 4 Mn^{2+} or 8 Ca^{2+}	
Pyruvate Carboxylase Chicken liver (mitochondria)	Synthesis of oxaloacetate Enzyme-Biotin + $MgATP^{2-}$ + HCO_3^- Acetyl-CoA ⇌ Enzyme-Biotin-CO_2 + $MgAOP^-$ + P_i^{2-} Enzyme-Biotin-CO_2 + pyruvate ⇌ Enzyme-Biotin + oxaloacetate	500 000 – 520 000	4 atoms of Mn^{2+}/ molecule of enzyme	1. Exists as a tetrameric unit which undergoes dissociation to inactive monomers at 0 °C 2. Protein has a Biotin : Manganese ratio of 1 : 1	9–12

Turkey liver			3.3 atoms of Mn^{2+}/ molecule of enzyme	1. Biotin : Manganese ratio of 1 : 1	13,14
Calf liver			1–1.5 atoms of Mn^{2+}/ molecule of enzyme	1. Biotin : Manganese ratio of 3 : 1 2. Biotin : Manganese + magnesium ratio of 1 : 1	
Diamine oxidase Human placenta and pregnancy plasma	$RCH_2NHR' + O_2 + H_2O \rightleftharpoons RCHO + H_2NR' + H_2O_2$	70 000	1.2 atoms of Mn^{2+}/ molecule of enzyme	1. Aggregation occurs at higher concentration 2. Protein also contains 1 atom of Cu^{2+}/molecule	15
Superoxide dismutase B-stearothermophilus	$O_2^{\overline{\cdot}} + O_2^{\overline{\cdot}} + 2H^+ \rightarrow O_2 + H_2O_2$	40 000	1 atom of Mn^{3+}/ molecule of enzyme	1. Native enzyme exists as a dimer of equal subunits (molecular weight 20 000 Daltons) 2. Each subunit composed of 185–190 amino acid residues 3. Resting state of manganese(III)	16–20
Escherichia coli		39 500	1.6–1.8 atoms of Mn^{3+}/molecule	a) Native enzyme exists as a dimer as a result of non-covalent interaction between subunits b) Native enzyme composed of 364 amino acid residues per dimeric unit c) Resting state of manganese(III) d) Replacement of manganese by other transition metal ions causes enzyme inactivation	21–23
Streptococcus mutans		40 000	2 atoms of Mn/ molecule of $S.O.D._I$	Native enzyme consists of two dimeric isoenzymes, SOD_I and SOD_{II}	24
Rhodopseudomonas spheroides		38 000	1 atom of Mn/ molecule of enzyme	1. Native dimeric enzyme consists of two identical subunits (molecular weight 18 400) non covalently linked	25

Table 3 (continued)

Protein and source	Function	Molecular weight (Daltons)	Manganese content	Properties	Ref.
				2. Dissociation of the dimer to monomeric form by treatment with Na-dodecylsulphate/2 mercapto-ethanol 3. Resting state of manganese(III)	
Pleutolus olearus		76 000–78 000	2 atoms of Mn/molecule of enzyme	1. Two superoxide dismutases found a) Minor S.O.D. (20%) composed of four identical subunits b) Major S.O.D. (80%) composed of two sets of subunits; α_2, β_2	26
Saccharomyces cerevisae		96 000	4 atoms of Mn/molecule of enzyme	1. Enzyme consists of four identical subunits 2. Dissociation of tetramer to dimer and monomeric form occurs upon treatment with Na-dodecyl sulphate/100 °C/mercapto ethanol 3. Enzyme located in mitochondrial matrix	27
Chicken liver		80 000	2.3 atoms of Mn/molecule of enzyme	1. Enzyme consists of four identical subunits covalently linked 2. Enzyme located in mitochondria	28
Human liver		85 300	3.9 ± .5 atoms of Mn/molecule of enzyme	Enzyme consists of four identical subunits (molecular weight 21 500)	29
Rat liver		500 000	1 atom of Mn/ $8-2.7 \times 10^5$ Daltons of protein	Enzyme consists of possibly four subunits (molecular weight 130 000 Daltons)	30

References for Table 3

1. Kalb, A.J., Lustig, A.: Biochim. Biophys. Acta *168*, 366 (1968)
2. Kalb, A.J., Levitzki, A.: Biochem. J. *109*, 669 (1968)
3. Agrawal, B.B., Goldstein, I.J.: Arch. Biochem. Biophys. *124*, 218 (1968)
4. Becker, J.W., et al.: J. Biol. Chem. *250*, 1513 (1975)
5. Cunningham, B.A., et al.: ibid. *250*, 1503 (1975)
6. Wang, J.L., et al.: ibid. *250*, 1490 (1975)
7. Sherry, A.D., Buck, A.E., Peterson, C.A.: Biochemistry *17*, 2169 (1978)
8. Paudolfino, E.R., Magnuson, J.A.: J. Biol. Chem. *255*, 870 (1980)
9. Scutton, M.C., Utter, M.F., Mildvan, A.S.: ibid. *241*, 3480 (1966)
10. Scrutton, M.C., Mildvan, A.S.: Biochemistry *7*, 1490 (1968)
11. Scrutton, M.C., Griminger, P., Wallace, J.C.: J. Biol. Chem. *247*, 3305 (1972)
12. Utter, M.F., Barden, R.E., Taylor, B.L.: Advan. Enzymol. *42*, 1 (1975)
13. Ochai, E.: Bioinorganic chemistry: An introduction. Allyn and Bacon, Inc., p. 388, 1977
14. Scrutton, M.C., Griminger, P., Wallace, J.C.: J. Biol. Chem. *247*, 3305 (1972)
15. Crabbe, M.J.C., et al.: Biochem. J. *155*, 679 (1976)
16. Harris, J.J., in: Superoxide and Superoxide Dismutases. (Eds. Michelson, A.M., McLord, J.M., Fridovich, I.), Academic Press, New York, p. 150, 1977
17. Bridgen, J., Harris, J.I., Northrop, F.: F.E.B.S. Lett. *49*, 392 (1975)
18. McAdam, M.E., Fox, R.A., Lavelle, F.: Biochem. J. *165*, 71 (1977)
19. Brock, C.J., Harris, J.I., Sato, S.: J. Mol. Biol. *107*, 175 (1976)
20. Brock, C.J., Harris, J.I.. Biochem. Soc. Trans. *5*, 1537 (1977)
21. Keele, B.B., McCord, J.M., Fridovich, I.: J. Biol. Chem. *245*, 22, 6176 (1970)
22. Ose, D.C., Fridovich, I.: ibid. *251*, 1217 (1976)
23. Fee, J.A.: see Ref. 16, p. 173
24. Varne, P.G., Keele, B.B., Rajagopalau, K.V.: J. Biol. Chem. *247*, 4782 (1972)
25. Lumsden, J., Cammack, R., Hall, D.O.: Biochim. Biophys. Acta *438*, 380 (1976)
26. Puget, K., Lavelle, F., Michelson, A.M.: see Ref. 16, p. 139
27. Ravindranceth, S.D., Fridovich, I.: J. Biol. Chem. *250*, 6107 (1975)
28. Weisiger, R.A., Fridovich, I.: ibid. *248*, 3582 (1973)
29. McCord, J.M., et al.: see Ref. 16, p. 128
30. McClure, W.R., Lardy, H.A., Kniefel, H.P.: J. Biol. Chem. *246*, 3569 (1971)

Table 4. Manganese proteins involved in photosynthetic processes

Protein source	Comments	Ref.
Zea mays L.	1. Molecular weight of 25 000 Daltons. 2. Contains 240 ng of manganese per mg of protein (1 Mn/8 molecules of protein). 3. Bound manganese is not displaced by tris-HCl (pH 8). 4. The manganese is thought to play a structural role, mainly in the stacking of the lamellae.	1
Spinacia oleracea L. (spinach)	1. Two soluble protein fractions were collected; a) molecular weight ⩾ 25 000 Daltons. b) molecular weight < 25 000 Daltons. 2. Manganese content of fractions a) 40 ng of manganese per mg of protein b) 240 ng of manganese per mg of protein 3. The fractions are thought to play a structural, rather than direct photosynthetic, role.	2
Spinacia oleracea (spinach) Lactuca sativa (lettuce) Phaseolus vulgaris (french beans) Hordeum vulgare (barley)	1. Three protein fractions of molecular weights 23 000, 26 000 and 29 000 Daltons collected. 2. Fractions contain 50–70 ng of manganese per gm of protein. 3. The manganese protein exhibits cyanide insensitive superoxide dismutase activity.	3
Phormidium luridium (algae)	1. Molecular weight of < 1000 Daltons. 2. Spheroplast have manganese content of 1 g atom Mn per 61 chlorophyll molecules. 3. The manganese protein increased oxygen evolution when reincorporated with fragmented spheroplasts (from 4 to 44 μmol-O_2-per mg chl per h).	4, 5
Phaeodactylum tricornutum (diatom)	1. Pigment-metallo-protein complex of molecular weight 850 000 Daltons consisting of a) Pigments, 40 molecules of chlorophyll-a 20 molecules of chlorophyll-c 20 molecules of fucoxanthin b) 40 protein subunits c) metal ions, 8 g atoms of copper 0.6–2.0 g atoms of manganese 2. The copper-manganese pigment-protein complex is thought to react in the early steps of the O_2-evolving system of photosystem-II.	6
Spinacia oleracea (spinach)	1. Molecular weight of 65 000 Daltons. 2. Two manganese atoms bound to each protein unit. 3. The bound manganese is removed by 0.8M Tris-HCl (pH 8.0). 4. The incorporation of the manganese protein to preformed liposomes restores up to 85% of oxygen evolving capacity. 5. The manganese protein is thought to be a part of the water-oxidizing enzyme system.	7

other metallo superoxide dismutases has recently been studied[167]. Recently, manganese superoxide dismutases have been detected in plant species including wheat germ[168], spinach leaves[169], Jerusalem artichoke[170], kidney beans[171], corn[172], mosses and ferms[173], peas[174] and tomato leaves[175]. Upon extraction of chloroplasts with organic solvents, bound manganese is retained with the aqueous protein phase. Cheniae and Martin[157] have shown that ^{54}Mn detected in whole chloroplast suspensions is associated with the aqueous protein material. Due to the unstable nature of the soluble protein, further characterization was not possible. Since then a number of workers have reported the isolation of proteins associated with photosynthetic oxygen evolution which are enriched with manganese (Table 4). Two groups of workers have previously reported the isolation of manganese protein complexes of similar molecular weights, 25,000 Daltons which differ greatly in manganese content (Table 4, Refs. 1, 2 respectively). It is thought that the heterogenous binding nature of the total manganese pool may be a contributing factor to the discrepancies[132]. The manganese associated with the complex is thought to play an important role in the structural organization of an active thylakoid membrane. Foyer and Hall have recently reported a rapid procedure for the preparation of light harvesting chlorophyll a/b protein complex from a variety of species including spinach, lettuce, french beans and barley (Table 4, Ref. 3). An assessment of the manganese content by atomic absorption after concentrated nitric acid digestion indicated a manganese content of 50–70 ng Mn/g-protein. The e.p.r. silent, bound manganese, could readily be removed by Tris–HCl, heat or freeze thawing to give free e.p.r. active manganese. There is a suggestion that the manganese may be involved in a superoxide dismutase function.

Huzisige et al.[176] first reported a protein which was directly involved in oxygen evolution. The protein, which was obtained by extraction of spinach chloroplasts treated with alkaline-Tris HCl buffer, was shown to have a marked effect on oxygen evolution in depleted chloroplasts. This oxygen evolution factor (OEF), as it was designated, was shown to accelerate oxygen production only under strong light. It was suggested that under such conditions another factor, designated X-factor, which together with the OEF induce oxygen production, is inactivated and that OEF functions to reactivate the inactive form of the X-factor.

The direct involvement of a manganese protein factor in oxygen evolution was demonstrated by Tel-Or et al. (Table 4, Refs. 4, 5). A low molecular, less than 1,000 D, water soluble factor was extracted from the algae *Phormidium luridium*.

References for Table 4

1. Lagoutte, B., Duranton, J.: F.E.B.S. Lett. *51*, 1, 21 (1975)
2. Henriques, F., Park, R. B.: Biochim. Biophys. Acta *430*, 312 (1976)
3. Foyer, C.H., Hall, D. O.: F.E.B.S. Lett. *101*, 2, 324 (1979)
4. Tel-Or, E., Stewart, W. D. P.: Nature *275*, 715 (1975)
5. Tel-Or, E., Avron, M.: Proceedings of the third international congress on photosynthesis, 1974. (Ed. Avron, M.) Elsevier, Amsterdam, p. 569, 1974.
6. Holdsworth, E.S., Arshad, J.H.: Arch. Biochem. Biophys. *183*, 361 (1977)
7. Spector, M., Winget, G. D.: Proc. Natl. Acad. Sci. U.S.A. *77*, 2, 957 (1980)

(a)

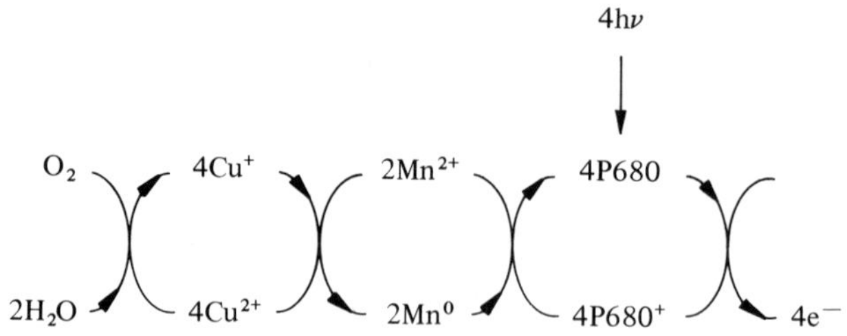

1.

(b)

$$M \xrightarrow{\oplus} M^+$$

$$(H_2O)^* \xrightarrow{\oplus} \underset{\displaystyle S_2}{(OH)^*} + H^+$$

$$(OH)^* + (H_2O)^* \xrightarrow{\oplus} \underset{\displaystyle S_3}{(H_2O_2)^{**}} + H^+$$

$$M^+ + (H_2O_2)^{**} \xrightarrow{\oplus} \{(O_2 \text{- -} H)^{**}_{\cdot\cdot} . . M^+ + H^+\}$$

$$\{(O_2 \text{- -} H)^{**}_{\cdot\cdot} . . M^+ + H^+\} \longrightarrow (O_2)^{**} + 2H^+ + M$$

$$(O_2)^{**} + 2H_2O \longrightarrow O_2 + (2H_2O)^*$$

2.

⊕; an oxidizing redox equivalent produced by $Chla_{II}$ and transferred to the water splitting enzyme (Y).

*,**; mono– and bi– nuclear complexation of functional manganese groups.

M; one electron donor.

(c)

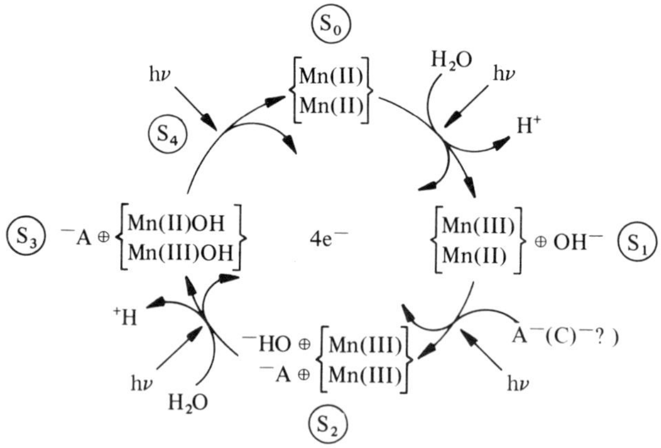

3.

Fig. 6a–c. Proposed models of oxygen evolution involving manganese

1. Holdsworth, E. S., Arshad, J. H.: Arch. Biochem. Biophys. *183*, 361 (1977)
2. Renger, G.: F.E.B.S. Lett. *81*, 223 (1977)
3. Wydrzynski, T., Sauer, K.: Biochim. Biophys. Acta *589*, 56 (1980)

Intact spheroplasts of *Phormidium luridium*, evolve O_2 but when a manganese containing (1 g-atom Mn/51 mol Chlorophyll) 'Hill factor' is released from broken spheroplasts, their O_2-evolving capacity is diminished. This latter capacity is restored when the factor is added back to the spheroplast preparations. In a recent review, however, Radmer and Cheniae [132] suggest that the oxygen evolution factor or Hill factor observed by Tel-Or may be some artifact.

The reported isolation of a Cu:Mn ($\approx 4:1$) pigment-protein complex from the photosystem II of *Phaeodactylum tricornutum*, a marine diatom, by Holdsworth et al. (Table 4, Ref. 6) led to a proposed mechanism of oxygen evolution which incorporated a manganese-copper redox couple in the water splitting reaction (Fig. 6a).

More recently, Spector and Winget (Table 4, Ref. 7) have reported the purification of a manganese containing protein extracted from spinach chloroplasts which was shown to activate oxygen evolution in reconstituted photosynthetic membranes. Fractionation of extracted proteins by Na-dodecylsulphate/polyacrylamide gel electrophoresis and subsequent chromatographic purification (Sephadex G-200/100) led to the isolation of an active protein with molecular weight of 65,000 Daltons. The presence of manganese in the protein was confirmed by use of spinach grown with ^{54}Mn supplementation, and the amount was determined by quantitative manganese analysis using atomic absorption spectroscopy which indicated that there were two manganese atoms bound to each 65,000 Daltons protein molecule. The ^{54}Mn could be released by the usual Tris treatment, leaving a manganese free apoprotein. Reconstruction of active membranes involved the addition of the manganese protein to depleted photosomes. These depleted photosomes were shown to catalyze high rates of photosystem I and II electron transport activity. After reconstitution with the 65,000 Daltons fraction the photosomes catalyzed high rates of methylamine uncoupled oxygen evolution. The ability of the 65,000 Daltons protein to restore Hill activity was lost when it was treated with Tris (pH 8.0), but similar treatment of photosomes prior to reconstitution with the protein had no effect. The results clearly indicate that restoration of oxygen-evolution capacity occurs as a result of the addition of a manganese-protein oxygen evolution factor.

3.4 Oxygen Evolution and Manganese

The effects of manganese depletion on the photochemical activity of photosystem II such as the quantum and flash yields of oxygen evolution, which are a measure of the fraction of the traps which are in the photoactive state and the abundance of the oxygen-evolving centres respectively, argue for a function of manganese at the oxidant site of photosystem II [128]. Removal of the loosely bound manganese pool, which comprises two-thirds of the total manganese content, by high concentrations of Tris–HCl (pH 8.0) results in a 90–100% loss of oxygen-evolving capacity. Between the limits of the remaining tightly bound manganese pool (one-third of the total manganese content) and the total manganese content, oxygen-evolution capacity in both strong and weak light was observed to be linear with manganese content. The role of the more tightly bound manganese pool has not yet been established, although some contribution towards lamellae stacking or superoxide dismutase activity is likely.

The removal of the loosely bound manganese pool does not inhibit the photo-oxidation of artificial electron donors or the electron transport processes of photosystems I and II. The mechanism of Tris-induced inactivation of oxygen evolution has recently been studied[177]. At concentrations of 0.8M Tris–HCl (pH 7.6 and 8.0) a slow rate of inactivation of $t_{1/2} = 20–25$ min occurs. The process could be accelerated ($t_{1/2} \approx 2.5$ min) by illumination with weak light of sufficient intensity to yield only 1 hit/System-II trap per 10 s and is rate limited by a dark step with a half-time of about 200 s. Chloroplasts inactivated with Tris in the dark could not be reactivated with either reduced dichloroindophenol ($DCIPH_2$) or the hydrazones FCCP (carbonyl-cyanide-*p*-trifluoromethoxyphenylhydrazone) and CCCP (carbonylcyanide-*m*-chlorophenylhydrazone) whereas illuminated chloroplasts when treated with such reagents were partially activated. As a result of the inactivation of 50–70% of the oxygen-evolving centres by one short saturating flash it was concluded that minimally one separation within system-II promotes a state which is susceptible to apparent inactivation by Tris and loss of oxygen-evolution capacity. It was proposed that as a result of a one quantum step, $S_1 \rightarrow S_2$, the S_2 state generated could co-ordinate through a $Mn^{(III)}$ ion to a Tris molecule via the amine function of Tris and resulting in a chemical and/or structural modification of S_2.

The inactivation of the oxygen-evolving enzyme by unprotonated hydroxylamine (Table 2, Ref. 5) and ammonia have previously been thought to occur as a result of the formation of a manganese-enzyme-amine state and in the case of ammonia co-ordination with the S_2 and S_3 states but not with the dark stable S_1-state[178,179]. Dissociation of bound manganese from the Tris modified S_2-state ($t_{1/2} \approx 200$ s) gives rise to inactive S-states (S_i). Reactivation of the S_i-states, by reinsertion of free manganese(II), is induced by reactivating reagents that cause the decay of S_2 or S_3 to lower S-states such as CCCP. The decay of the potentially activable state ($t_{1/2} \approx 175$ min) is correlated with diffusion of the unbound larger two-thirds manganese pool of System-II within the thylakoid into the suspending medium.

The loosely bound two-thirds manganese pool which is required was estimated to constitute 4–6 manganese atoms per oxygen evolving centre, although the isolation of a manganese oxygen evolution protein containing 2 manganese atoms has added supportive evidence to recently proposed models of oxygen evolution involving manganese[128].

3.5 *Models of Oxygen Evolution Involving Manganese*

Based on earlier estimates of four manganese atoms per active Photosystem-II, Govindjee et al.[180] proposed a reaction scheme involving four manganese atoms which undergo oxidation-reduction changes between manganese(II) and manganese(III) states as the enzyme cycles through the S-states. In more recent investigations binuclear manganese protein complexes have become the active centres in models which illustrate the four electron photo-oxidation of water (Fig. 6a–c). Holdsworth et al. have proposed a mechanism which involves a $2\,Mn^0/Mn^{2+}$ couple to accommodate the four electrons derived from water which are then transferred to the photo-oxidized state of P680 (Fig. 6a). Included in the model is an attempt to assign a role to the copper found in the pigment-protein complex isolated and which has previously been outlined in Table 4.

Electron paramagnetic resonance studies showed that upon illumination re-oxidation of Mn^0 (or Mn^I) and Cu^{1+} to Mn^{2+} and Cu^{2+} occurred.

The model proposed by Renger (Fig. 6b) assumes that the reactions occur in the water splitting enzyme unit Y or Z whose properties could be described by three postulates:

1. The reactive intermediates of the sequential one step water oxidation to oxygen are assumed to be stabilized at functional groups which contain Mn as the central ion.
2. Oxygen is ultimately formed via the oxidation of complexed superoxide by a 1 electron donor M, whose oxidized form M^+ is stable in the dark. The complexed oxygen is spontaneously released by an exergonic ligand-ligand exchange with two water molecules.
3. The preformation of the oxygen bond occurs at the level of hydrogen peroxide, which is binuclearly complexed by two functional manganese groups.

The complexation of hydrogen peroxide, superoxide and oxygen [181,182] to manganese is thought to occur via direct interaction with metal ion 3 d-orbitals. Additionally it is postulated that a charge delocalization between the central manganese ion and the corresponding ligand is essential for complexation. The structural organization of the water splitting unit is described as containing two manganese atoms at a distance short enough to allow binuclear complexation for the peroxide and superoxide intermediates which are ordered around electron carriers D_1 and M. D_1 acts to connect the water splitting enzyme Y and the oxidizing equivalent generated by P680-Chla_{II}. The ground state of manganese is thought to be + 2 or + 3.

The most recent attempt to deduce the mechanism of oxygen evolution has been reported by Sauer and Wydrzynski (Fig. 6c). Measurements of oxygen evolution and proton release together with e.p.r. analysis of the manganese of heat treated chloroplasts led to the proposed scheme as outlined in Fig. 6c. The scheme illustrates the partial oxidation of water prior to oxygen evolution and the complete oxidation of two functional manganese(II) ions to the higher manganese(III) state. Charges are balanced in the $S_0 \rightarrow S_1 \rightarrow S_2$ transition by a hydroxyl and an inorganic anion, possibly Cl^-. Kelley et al. [183] have previously suggested chloride ion can bind to a manganese protein complex. Complexation of a second OH^- ($S_2 \rightarrow S_3$) results in a transfer of electrons to the manganese atoms, thus decreasing their oxidation level and increasing that of the bound oxygens to the level of peroxide. The final step ($S_3 \rightarrow S_4 \rightarrow S_0$) forms molecular oxygen by removing two more electrons from the oxygen atoms, releasing the remaining two protons and returning the two manganese atoms to the manganese(II) state. Although higher oxidation states or larger complexes of manganese are not completely discounted, the proposed scheme illustrates that only manganese(II) and manganese(III) states need to be involved in the stabilized S-state and that water splitting reactions involve hydroxyl ligand partial oxidation in a manganese-protein complex. With the advent of the recent isolation of a manganese-protein complex, further investigations may resolve the level of contribution of manganese to the S-states. Other workers are studying *in vivo* manganese contributions by synchroton radiation techniques, although no results have yet become available[54].

The essential component of the models proposed is the nature of the redox state (S) of manganese. Both electron and nuclear paramagnetic resonance have been employed in the study of chloroplast manganese.

3.6 EPR Studies of Chloroplast Manganese

Investigation into the functional state of manganese by the e.p.r. method in a number of cases is made difficult by the fact that the complexing of the ion with low and high molecular weight compounds leads to the disappearance of the characteristic six-component e.p.r. spectrum, because of the heavy widening due to the splitting of the spin sublevels in the zero field on disturbance of the cubic symmetry of the field about the ion[184].

E.p.r. signals from hydrated manganese(II) ions with $g = 2.003$ and $A \cong 94.5$ G in photosynthetic systems are observable at room temperature in highly native specimens such as leaves, algae and chloroplasts[31,185]. The e.p.r. detectable manganese is observed by treatment of the cell chloroplasts with conditions leading to displacement of the bound manganese, as manganese(II) or higher valent e.p.r. non-detectable manganese, from the thylakoid into the cell matrix and loss of water splitting enzyme activity. E.p.r. detectable manganese(II) has been observed in chloroplast cells treated with Tris–HCl [160,161], high concentrations of divalent cations[184,186], reductants[184,187] and mild thermal shock[132,188]. A detailed account of the e.p.r. aspects of photosynthesis including the manganese contribution can be found in a recent review[31].

Gribova and Tikhonov[184] have observed e.p.r. spectra of Mn^{2+}(I) in chloroplasts isolated in salt free or salt buffers which significantly differs in peak to peak separation and relative intensity, from that of the hydrated ion, $Mn(H_2O)_6^{2+}$(II). The transition from I⇌II in a number of cases is induced by light in the presence of reductants.

Blankenship and Sauer[160,161] found that the e.p.r. of free Mn(II) extracted from Tris-treated chloroplasts was light sensitive; decaying on illumination and recovering in the dark. Lozier et al.[189] had previously observed a similar result with *Chlorella pyrenoidosa* treated with chaotropic agents. They concluded that Tris washing releases part of the bound manganese to the interior space of the thylakoid and that the uptake of manganese from the external medium is a light driven reaction, but once within the thylakoid, the divalent manganese is incorporated into the membrane in the dark with resulting loss of e.p.r. signal and reactivation of oxygen evolution.

Recently Siderer et al.[190] have observed an e.p.r. signal of manganese(II) ions in *untreated* lettuce chloroplasts at room temperature. The spectrum is characterized by a spin Hamiltonian,

$$\mathcal{H} = g\beta H \cdot S + AI \cdot S$$

with the following parameters: $g = 2.003 \pm .001$; hyperfine splitting, $A = 94.5 \pm 0.5$ G and an average peak to peak linewidth of $\Delta H \cong 30$ G. Intensity measurements were performed by comparing the intensities of the manganese(II) signals of two identical chloroplast preparations, one of which was slightly acidified. The authors obtained a ratio of control/acidified ~ 0.2. This result is close to the expected relative intensity of the $-1/2 \rightarrow 1/2$ transitions and suggests that the observed signal in the control chloroplast preparation comes from $M_s = 1/2$ transitions of bound manganese, rather than from free manganese(II) ions. A detailed analysis of the e.p.r. (Q and X band) spectra of single crystals and aqueous solutions of manganese proteins has recently been reported[191,192]. These results show that only the $-1/2 \leftrightarrow 1/2$ transition of the Mn^{2+} spectra are observed in concanavalin A solutions, corresponding to 0.31 of the total e.p.r. intensity in good agreement with the 0.26 expected if only $M = 1/2$ transitions are observed in the manganese(II) protein complex. E.p.r., in conjunction with n.m.r. (^{13}C, ^{31}P and ^{1}H), techniques using manganese(II) as a paramagnetic probe have been extensively used in the study of manganese protein binding sites in a range of protein sources[193–212]. Of particular interest has been the investigation of regulatory and substrate binding sites, and the distances separating regulatory sites and regulatory sites from substrate sites, of enzymes. Recently the distance between the two metal ion binding sites designated n_1 and n_2 of glutamine synthetase derived from *Escherichia coli*, which catalyzes the reaction

$$\text{L-glutamate} + \text{ATP} + \text{NH}_3 \xrightleftharpoons{2M^{2+}} \text{L-glutamine} + \text{ADP} + \text{Pi}$$

was deduced from the paramagnetic interaction between manganese(II) nucleotides bound at n_2 and chromium(III) bound at site n_1. A decrease in spectral intensity of 66% of the bound manganese due to chromium(III) was correlated to a separation of n_1 and n_2 by 7 ± 2 Å[209].

In an earlier study on normal chloroplast preparation from *Chlorelia pyrenoidosa*, Treharne et al.[213] had observed a manganese(II) e.p.r. signal which was found to increase in intensity in the dark whilst diminishing when illuminated. A free radical spin signal at $g = 2$ was observed to increase and decrease in intensity in the opposite fashion to the manganese(II) signal and at at much faster rate.

Siderer et al. found that illumination of the chloroplasts in the cavity by visible light causes a reduction in the intensity of the manganese(II) signal by about 80% in one example. Turning the light off results in restoration of the signal to its original intensity. A free radical signal at $g = 2.0$ at the centre of the manganese multiplet which is generally referred to as signal I and signal II, representing the oxidized form of P700 and a radical connected with system II, respectively was found to behave similarly to that observed by Treharne[213] upon alternate light/dark periods.

A number of treatments and reagents including heat, Tris washing, magnesium cations, DCMU, artificial electron donors and acceptors and FCCP (carbonylcyanide-p-trifluoromethoxyphenylhydrazone), were used to study the relevance of the light-induced change of the manganese(II) signal to chloroplasts' electron transport activity. DCMU, which blocks electron transfer from PS II and hence the manganese site to plastoquinone, was found to inhibit the light effect on the manganese(II) signal. Heat treatment, Tris washing and incubation with magnesium cations resulted in an increase of the dark manganese(II) signal apparently due to release of part of the bound manganese. The resulting signal was still affected by light. However, the kinetics of both light and dark reactions were considerably slower after the treatments. The artificial electron donor phenylenediamine affects the light response of the manganese(II) signal which the authors suggested is most probably due to the reduction of the oxidized manganese (e.p.r. silent) form to manganese(II) by phenylenediamine. In the presence of ascorbate (1mM), a 2 electron donor, the light effect on the manganese(II) signal was completely abolished. FCCP was shown to decrease the light response of the manganese(II) signal and at around 3×10^{-4}M completely inhibits the reaction. FCCP has a two-fold role (i) as an uncoupler, and (ii) as an ADRY reagent (accelerating the deactivation of the reaction of the water splitting enzyme (Y)). There are three main points that emerge from the above results:

1. Manganese(II) ions participate in the electron transfer process of the photosynthetic cycle via oxidation-reduction equilibria involving +3 or higher oxidation states of manganese.
2. The active manganous ions are bound in an asymmetric environment within the chloroplast structure, apparently in protein complexes.
3. The effects of the various reagents on the light response of the manganous signal suggest that its photo-oxidation is associated with photosystem II, the most probable site being the water splitting apparatus.

Recently Siderer and Malkin[214] have extended the e.p.r. studies on flash-induced manganese(II) oxidation in lettuce chloroplasts. A comparison of the amplitudes of the manganese(II) e.s.r. signal shows that in a single saturating flash only ~ one-eighth of the manganese which reacted in continuous light was oxidized. Estimation of the manganese concentration in photosystem II gives numbers in the vicinity of 8 atoms/reaction centre. Thus there is a single manganese(II) oxidation corresponding to each turnover of the reaction centre, and there is a pool of ~ 8 manganese(II) ions which are oxidized only by continuous light. In a series of 4 flashes, 2.5 S apart with 52 s dark periods before the repetition of the experiment, the maximal decrease in the amplitude of the manganese signal occurred after the third flash. Noise levels prevented a clear assessment of the signal reduction after the fourth flash and hence a correlation between the photo-oxidation pattern of manganese by light flashes and the postulated oxidation changes of manganese(II) in the S-states as proposed by Kok et al.[136].

Wydrzynski and Sauer[215] have recently reported on the periodic changes in the oxidation state of manganese in spinach chloroplasts upon illumination with flashes. The authors investigated the manganese(II) e.p.r. signal in heat treated chloroplasts in the presence of oxidants and reductants before and after illumination, either continuous or as a series of brief flashes. By using the peak to trough height of the manganese(II) hexaaqua signal as a measure of the relative concentration of detectable manganese released upon heat treatment, the authors had found that pre-illuminated fresh chloroplast preparations released up to 43% less e.p.r. detectable manganese(II) than preparations kept under dark conditions. The addition of DCMU and FCCP during the pre-illumination period eliminates any light induced change in the amount of detectable manganese(II) release. These results implicate the involvement of photosystem II reactions in the light effect on manganese release. In order to determine whether photooxidation of some bound manganese(II) to e.p.r. non-detectable higher oxidation states or a decrease in the amount of manganese(II) released as a result of some macro-conformational change is responsible for the reduction in the intensity of the $Mn(II)(H_2O)_6^{2+}$ signal; the chloroplasts were subjected to glutaraldehyde fixation, which prevents possible macro-conformational changes. Pre-illumination of fixed chloroplasts again exhibited the observed light effect indicating that the chloroplasts release both detectable manganese(II) and non-e.p.r.-detectable higher valent state manganese upon heat treatment.

This high valent manganese is thought to be Mn(III) which undergoes disproportionation in aqueous media as follows: 2 Mn(III) $\rightarrow$ Mn(II) + Mn(IV). The addition of H_2O_2 at pH 6.0, but not pH 7.5, to pre-illuminated chloroplasts resulted in an increase in detectable manganese(II), presumably as a result of manganese(IV) reduction to manganese(II). The pH dependance of the hydrogen peroxide reaction suggests that the non-e.p.r.-detectable form of the manganese is MnO_2. The light induced change in manganese release is consistent with a net photo-oxidation of the bound manganese.

3.7 Proton Relaxation studies of Chloroplast Manganese

Proton relaxation techniques using manganese(II) as a paramagnetic probe have been applied to elucidate the environment of protein binding sites[216)].

In recent years measurements of the enhancement of the longitudinal relaxation rate, $1/T_1$, of water protons by manganese(II)[201,202,204,206,210,211)] have been applied to the investigation of chloroplast bound manganese. Manganese(II) usually maintains some water ligands when bound to large macromolecules and the electronic relaxation of manganese(II) is such that the relaxation rates of the protons in the solvation sphere are strongly enhanced. Preliminary experiments showed that treatments which affect the amount of bound manganese in chloroplast membranes have a large effect on the spin-lattice or longitudinal relaxation rate, $1/T_1$ of water protons[217)]. The addition of reductants such as tetraphenylboron and hydroxylamine caused a large increase in $1/T_1$, whereas the addition of oxidants (potassium ferricyanide or 2,6-dichlorophenolindophenol) led to a decrease in this rate, suggesting that proton relaxation rates (PRR) could be used to detect changes in the oxidation state of manganese. It was suggested that manganese exists as a mixture of oxidation states, possibly Mn(II) and Mn(III), in dark-adapted chloroplasts. In a recent set of experiments the aqueous suspensions were exposed to a series of intense light flashes after n of which ($n = 0, 1, 2 \ldots 20$) the proton transverse (spin-spin) relaxation rate, $1/T_2$, was measured[218)]. The response of $1/T_2$ to light flashes was found to be very similar to that of oxygen evolution, both sequences exhibiting a damped oscillation with a periodicity of four flashes. The oscillatory pattern of $1/T_2$ does however have some significant differences which the authors and others[132)] have outlined. Minima in the relaxation rates for example occur after the 4th, 8th, 12th, etc. flashes whilst minima in the oxygen yield occur after the 6th and 10th. It is thought that the differences in $1/T_2$ and oxygen yield patterns can be attributed to the fact that the relaxation rates may differ significantly for each of the various S-states, whereas oxygen evolution only takes place during the $S_4 \rightarrow S_0$ transition. Some caution has been expressed by Lawrence et al. [181] on the interpretation of the proton relaxation effects since these rates may be sensitive to many factors other than manganese.

In a more recent investigation of proton and oxygen-17 nuclear magnetic resonance relaxation rates (T_1^{-1} and T_2^{-1}) of aqueous suspensions of dark adapted chloroplasts, it was concluded that T_1^{-1} and T_2^{-1} are determined largely by the loosely bound manganese present in the chloroplast membranes [219]. Gol'dfel'd et al. [31, 220] report similar findings and conclude that the contribution of bound manganese depends on the state of the chloroplast, in particular, the state of the system of photo-oxidation of water. Wydrzynski et al. [219] found that the effects of oxidants (e.g. ferricyanide) and reductants (e.g. tetraphenylboron) on the proton relaxation rate indicate that only about one-third to one-fourth of the loosely bound manganese is present in the dark adapted chloroplasts as manganese(II), the remainder being in a higher oxidation state (S) probably manganese(III).

3.8 Effect of Chloride Ion on Oxygen Evolution

Firm evidence for a specific function of chloride in reactions closely associated with oxygen evolution was first provided by Izawa [221]. The addition of chloride ($\sim$5mM) resulted in a 4- to 10-fold stimulation of oxygen evolution of chloroplasts depleted of chloride and uncoupled by EDTA. Alternatively it is thought that chloride might protect against photoinactivation by enhancing electron transport between P680 and H_2O oxidation, thereby removing the deleterious oxidants of system II [132].

The interaction of chloride with manganese(II) has been implicated by recent e.p.r. studies of a number of plant species including cypress needles [222]. The e.p.r. spectra show the typical sextet pattern for high spin manganese(II) centred at $g \approx 2.02$. The signal also shows quartet splitting arising from interaction with nuclei of $I = 3/2$. Both chloride and copper(II) have $I = 3/2$, however the observed quartet splitting is thought to derive from a Mn–Cl interaction. Chloride replacement by fluoride ($I = 1/2$) resulted in doublet splitting thus confirming a Mn-halide linkage. Gol'dfel'd and Blumfeld [31] suggest that the Mn–Cl complex in photosynthetic membranes is a product of the partial catabolism of the water-splitting active structure, which can accumulate in sufficient quantity only in certain species. The manganese(II)-chloride complex is thought not to be identical with the water-splitting complex itself, however this end product of degradation preserves some features of the initial structure and indicates the presence of the Mn–Cl link.

Recently Kelley and Izawa [183] have measured the rate of oxygen evolution (using silicomolybdate and 2,5-dimethyl-*p*-benzoquinone (DMQ) in tobacco chloroplasts and ferricyanide in spinach chloroplasts as artificial electron donors) of EDTA uncoupled, functionally chloride depleted chloroplasts in the presence of several inorganic and organic anions. The order of effectiveness of the anions to restore oxygen evolving capacity was found to be as follows:

$$Cl^- > Br^- > NO_3^- \geqslant I^- \geqslant HCOO^- \geqslant HCO_3^- > F^- \text{ (no effect)}.$$

Acetate was found to be totally ineffective as were other di- and trivalent anions such as citrate, oxalate, succinate and tartate in activating oxygen evolution capacity of chloroplasts depleted of anions. It was suggested that the site of chloride involvement is on the water side of PS II and the step requiring chloride is one which is specifically associated with the water-splitting mechanism. The authors dismiss the possibility that the role in water oxidation is to act as a counter ion to enable the internal release of protons as the model proposed by Wydrzynski et al. Fig. 6c suggests, since the photooxidation of hydrogen peroxide by PS II does not require chloride. Alternatively a proposal was put forward which suggested that chloride is a cofactor of an enzyme involved in oxygen evolution and that it plays a prominent role in preventing enzyme deactivation by hydroxylamine by binding to the manganese enzyme, and that this ligand binding somehow alters the susceptibility of the enzyme to hydroxylamine attack.

Recent investigations have shown that chloride depleted cells treated with exogenous manganese(II) show a 50% reduction in Hill activity which can achieve 100% reduction when illuminated with red light[223]. The addition of chloride at concentrations less than ~ 1mM to chloroplasts prior to manganese(II) treatment prevents loss of Hill activity by a process which does not involve complexation of the free manganese(II) ions. The exogenous manganese(II) inhibits the oxygen evolving mechanism but does not affect the reaction centre of photosystem II. Furthermore the manganese(II) sensitive step and the chloride requiring step are closely related to each other. It is proposed that chloride binds directly to a chloride free temporarily non-functional enzyme, E_f, which is involved in oxygen evolution:

$$E_f + Cl^- \rightleftharpoons E \cdot Cl^- .$$

Inhibition by exogenous manganese(II) is thought to occur as a result of attack of E_f, but not the chloride enriched form, by manganese(II). In order to explain the 50% inhibition of dark adapted chloroplasts it is suggested that the chloride deficient enzyme exists in two different oxidation states (about 50% of each state), $E_{f.red}$ and $E_{f.ox}$, and only the more oxidized form, $E_{f.ox}$, is sensitive to manganese(II). The inhibition is thought to involve reduction of the oxidized form by manganese(II) which is irreversible and does not allow attack and reactivation by chloride ions.

4 References

1. Hall, D. O.: Solar energy *22*, 307 (1979)
2. Hall, D. O., in: Research in photobiology (Ed. by Castellani, A.). Plenum Press, N.Y., p. 347, 1977
3. Hall, D. O.: Fuel *57*, 322 (1978)
4. Burns, R.H., Black, C.C. (Eds.): CO_2 metabolism and plant productivity, University Park Press, 1976
5. Gibbs, M., Latzko, E. (Eds.): Encyclopedia of plant physiology, New Series 6: (Photosynthetic carbon metabolism and related processes), Springer, Berlin, Heidelberg, New York, 1979

6. Hall, D.O., Rao, K.K. (Eds.): Photosynthesis, London, 1977
7. Mühlethaler, K., in: Encyclopedia of plant physiology. New Series 5: (Photosynthesis I: photosynthetic electron transport and photophosphorylation (Eds. Trebst, A., Avron, M.)), p. 503, 1979
8. Sane, P.V.: see Ref. 7, p. 522
9. Miller, R.K., in: Development in plant biology, 2 (Chloroplast development (Eds. Akogunoglou, G., Argyroudi-Akogunoglou, J.M.), Elsevier/North Holland, p. 17, 1978
10. Gregory, R.P.F.: Biochem. Soc. Transactions *7*, 1222 (1979)
11. Vernon, L.P., et al.: Photochem. Photobiol. *14*, 343 (1971)
12. Butler, W.L., in: Ciba foundation symposium 61, New Series. (Chlorophyll organization and energy transfer in photosynthesis) Excerpta Medica, Amsterdam, p. 237, 1979
13. Butler, W.L., Kitajima, M.: Biochim. Biophys. Acta *396*, 72 (1975)
14. Kitajima, M., Butler, W.L.: ibid. *408*, 297 (1975)
15. Lieberman, J.R., Bose, S., Arntzen, C.J.: Biochim. Biophys. Acta *502*, 417 (1978)
16. Govindjee, et al.: Z. Naturforsch. *346*, 826 (1979)
17. Junge, W.: see Ref. 7, p. 59
18. Katz, J.J., Shipman, L.L., Norris, J.R.: see Ref. 12, p. 1
19. Katz, J.J., et al.: Ann. Rev. Biophys. Bioenerg. *7*, 393 (1978)
20. Thornber, J.P., Barber, J.: Topics in Photosynthesis 3 (Photosynthetic related model systems) p. 27, 1979
21. Sauer, K.: Acc. Chem. Res. *11*, 257 (1978)
22. Pellin, M.J., Kaufman, K.J.: Nature *278*, 54 (1979)
23. Dougherty, et al.: J. Am. Chem. Soc. *102*, 1, 416 (1980)
24. Brace, J.G., et al.: ibid. *100*, 16, 5203 (1978)
25. Boxer, S.G., Bucks, R.R.: ibid. *101*, 7, 1883 (1979)
26. Fong, F.K., Koester, V.J., Galloway, L.: ibid. *99*, 7, 2372 (1977)
27. Fong, F.K., Wassam, W.A.: ibid. *99*, 2375 (1977)
28. Sarai, A.: Biochim. Biophys. Acta *589*, 71 (1980)
29. Davis, M.S., Forman, A., Fajer, J.: Proc. Natl. Acad. Sci. U.S.A. *70*, 4170 (1979)
30. Warden, J.T., in: Biological Magnetic Resonance I (Eds. Berliner, L.J., Reuben, J.) Plenum Press, New York, p. 239, 1978
31. Gol'dfel'd, M.G., Blumfeld, L.A.: Bull. Magn. Res. *1*, 2, 66 (1979)
32. Weaver, E.C., Corker, G.A.: see Ref. 7, p. 168
33. Malkin, R., Bearden, A.J.: Biochim. Biophys. Acta *396*, 250 (1975)
34. Visser, J.W.M., Rijgersberg, C.P., Gast, P.: ibid. *460*, 36 (1977)
35. Valenty, S.J., Gaines, G.L. Jr.: J. Am. Chem. Soc. *99*, 1285 (1977)
36. Sprintschnik, G., et al.: J. Am. Chem. Soc. *99*, 4947 (1977)
37. Gaines, G.L. Jr., Behnken, P.E., Valenty, S.J.: ibid. *100*, 21, 6549 (1978)
38. Valenty, S.J.: ibid. *101*, 1 (1979)
39. Janzen, A.F., Bolton, J.R., Stillmann, M.J.: ibid. *101*, 6337 (1979)
40. Mercier-Smith, J.A., Whitten, D.G.: ibid. *101*, 6620 (1979)
41. Heither, H., Galla, H.J., Möhwald, H.: Z. Naturforsch. *33 C*, 382 (1978)
42. Schönfeld, M., Montal, M., Feher, G.: Proc. Natl. Acad. Sci. U.S.A. *76*, 6351 (1979)
43. Seefeld, K.P., Mobius, D., Kuhn, H.: Helv. Chim. Acta *60*, 2068 (1977)
44. Young, R.C., Feldberg, S.W.: Biophys. J. *27*, 237 (1979)
45. Kraljic, I., Barboy, N., Leicknamm, J.P.: Photochem. Photobiol. *30*, 631 (1979)
46. Menger, F.M.: Acc. Chem. Res. *12*, 111 (1979)
47. Grätzel, M., in: Micellization, solublization, and microemulsions (Ed. Mittal, K.C.) Plenum Press: New York, p. 531, 1977
48. Thomas, J.K., Almgren, M., Grieser, F.: J. Am. Chem. Soc. *101*, 2021 (1979)
49. Thomas, J.K., Piciulo, P.L.: ibid. *100*, 3239 (1978)
50. Grätzel, M., Thomas, J.K., in: Modern Fluorescence Spectroscopy 2 (Ed. by Wehry, E.L.) Plenum Press, New York, p. 196, 1976
51. Hall, G.E.: J. Am. Chem. Soc. *100*, 8262 (1978)
52. Thomas, J.K., Piciulo, P.L.: ibid. *101*, 2502 (1979)

53. Van-Ginkel, G.: Photochem. Photobiol. *30*, 397 (1979)
54. Calvin, M.: Acc. Chem. Res. *11*, 369, (1978)
55. Humphry-Baker, R., et al.: J. Am. Chem. Soc. *102*, 3689 (1980)
56. Brugger, P.A., Grätzel, M.: ibid. *102*, 2461 (1980)
57. Luisetti, J., Möhwald, H., Galla, H.J.: Z. Naturforsch. *34C*, 406 (1979)
58. Ford, W.E., Otvos, J.W., Calvin, M.: Nature *274*, 507 (1978)
59. Escabi-Perez, J.R., et al.: J. Am. Chem. Soc. *101*, 2231 (1979)
60. Infelta, P.P., Grätzel, M., Fendler, J.H.: ibid. *102*, 1479 (1980)
61. Nomura, T., et al.: ibid. *102*, 1485 (1980)
62. Futami, A., Hurt, E., Hauska, G.. Biochim. Biophys. Acta *547*, 583 (1979)
63. Futami, A., Hauska, G.: ibid. *547*, 597 (1979)
64. Stillwell, W., Tien, H.T.: Biochem. Biophys. Res. Comm. *81*, 212 (1978)
65. Toyoshima, Y., et al.: Nature *265*, 187 (1977)
66. Monserrat, K., Grätzel, M., Tundon, P.: J. Am. Chem. Soc. *102*, 5137 (1980)
67. Kiwi, J., Grätzel, M.: ibid. *100*, 6314 (1978)
68. Kiwi, J., Grätzel, M.: Angew. Chem. Int. Ed. Engl. *18*, 624 (1979)
69. Boardman, N.K.: Ann. Rev. Plant Physiol. *21*, 115 (1970)
70. Avron, M., in: Structure and function of chloroplasts (Ed. Gibbs, H.), Springer, Berlin, Heidelberg, New York, p. 149, 1971
71. Bishop, D.G.: Photochem. Photobiol. *20*, 281 (1974)
72. Avron, M. (Ed.): Proc. Third Internat. Congr., Photosynthesis I (Primary reactions and electrons transport) Elsevier, Amsterdam, 1975
73. Bearden, A.J., Malkin, R.: Q. Rev. Biophys. *7*, 131 (1975)
74. Williams, W.P., in: Topics in Photosynthesis, 2 (Primary processes of photosynthesis (Ed. by Barber, J.) Elsevier, Amsterdam, p. 99, 1977
75. Golbeck, J.H., Lien, S., San Pietro, A.: see Ref. 7, p. 94
76. Bendall, D.S.: Biochem. Soc. Trans. *6*, 372 (1978)
77. Bearden, A.J., Malkin, R.: Curr. Top. Bioenerg. *7*, 139 (1978)
78. Cramer, W.A.: see Ref. 7, p. 227
79. Knaff, D.B.: Co-ordn. Chem. Revs. *26*, 47 (1978)
80. Rich, P.R., Bendall, D.S.: Biochim. Biophys. Acta *591*, 153 (1980)
81. Horton, P., Croze, E.: ibid. *462*, 86 (1977)
82. Horton, P., Croze, E., Smetzer, G.: ibid. *503*, 274 (1978)
83. Hoffmann, P., Huke, B., Kockntz, A.: Biochem. Physiol. Pflanz (BPP), 579 (1979)
84. Izawa, S.: see Ref. 7, p. 266
85. Sauer, K.: Ann. Rev. Phys. Chem. *30*, 155 (1979)
85a. Klimov, V.V., et al.: F.E.B.S. Lett. *82*, 183 (1977)
86. Fujita, I., Davis, M.S., Fajer, J.: J. Am. Chem. Soc. *100*, 6280 (1978)
86a. Klimov, V.V., Ke, B., Dolan, E.: F.E.B.S. Lett. *118*, 123 (1980)
87. Amesz, J.: see Ref. 7, 238
88. Ulrich, E.L., Markley, J.L.: Co-ordn. Chem. Revs. *27*, 109 (1978)
89. Katoh, S.: see Ref. 7, p. 246
90. Boas, J.F., Pilbrow, J.R., Smith, T.D.: see Ref. 28, p. 277
90a. Colman, P.M., et al.: Nature *272*, 319 (1978)
91. Cookson, D.J., Hayes, M.T., Wright, P.E.: ibid. *283*, 682 (1980)
92. Cookson, D.J., Hayes, M.T., Wright, P.E.: Biochim. Biophys. Acta *591*, 162 (1980)
93. Segal, M.G., Sykes, A.G.: J. Am. Chem. Soc. *100*, 4585 (1978)
94. Rieske, J.S., Hansen, R.E., Zaugg, W.S.: J. Biol. Chem. *239*, 3017 (1964)
95. Rieske, J.S., Zaugg, W.S., Hansen, R.E.: ibid. 239, 3023 (1964)
96. Malkin, R., Aparico, P.J.: Biochem. Biophys. Res. Commun. *63*, 1157 (1975)
97. Malkin, R., Bearden, A.J.: Co-ordination Chem. Rev. *28*, 1 (1979)
98. Malkin, R., Bearden, A.J.: Biochim. Biophys. Acta *505*, 147 (1978)
99. Reimer, S., Link, K., Trebst, A.: Z. Naturforsch. *34C*, 419 (1979)
100. Oettmeier, W., Reimer, S., Link, K.: ibid. *33C*, 695 (1978)
101. Bayley, H.: ibid. *34C*, 490 (1979)

102. Trebst, A., et al.: ibid. *34C*, 831 (1979)
103. White, C.C., Chain, R.K., Malkin, R.: Biochim. Biophys. Acta *502*, 127 (1978)
104. Malkin, R., Posner, H.B.: ibid. *501*, 552 (1978)
105. Chain, R.K., Malkin, R.: Arch. Biochem. Biophys. *197* (1) 52 (1979)
106. Ereciriska, M., Wilson, D.F.: ibid. *102*, 80 (1979)
107. Dismukes, G.C., Sauer, K.: Biochim. Biophys. Acta *504*, 431 (1978)
108. Sauer, K., et al., in: Bioenergetics of membranes (Eds. Packer, L., Papageorgiou, G.C., Trebst, A.) p. 351, 1977
109. Sauer, K., et al.: Biochim. Biophys. Acta *503*, 120 (1978)
110. Shuvalov, V.A., Dolan, E., Ke, B.: Proc. Natl. Acad. Sci. U.S.A. *76*, 2, 770 (1979)
111. Baltimore, B.G., Malkin, R.: Photochem. Photobiol. *31*, 485 (1980)
112. Adams, M.W.W., et al.: Biochim. Biophys. Res. Commun. *79*, 1184 (1977)
113. Adams, M.W.W., et al.: Biochim. Biophys. Acta *589*, 1 (1980)
114. Lovenberg, W. (Ed.): The Iron-Sulfur Proteins, Vol. 1 & 2, Academic Press, New York, 1973
115. Lovenberg, W. (Ed.): ibid. Vol. 3, Academic Press, New York, 1977
116. Palmer, G.. The Enzymes, Vol. 12 (Ed. by Boyer, P.), 3rd ed. Part B, Academic Press, New York, p. 2, 1975
117. Hall, D.O., Cammack, R., Rao, K.K., in: Iron in Biochemistry and Medicine (Eds. Jacobs, A., Woodwood) Academic Press, p. 279, 1974
118. Smith, T.D., Pilbrow, J.R., in: Biological Magnetic Resonance 2 (Eds. Berliner, L.J., Reuben, J.) Plenum Press, New York, p. 85, 1980
118a. Sweeney, W.V., Rabinowitz, J.C.: Ann. Rev. Biochem. *49*, 139 (1980)
119. Haynk, B.H., et al.: J. Biol. Chem. *255*, 8, 3242 (1980)
120. Stout, C.D., et al.: ibid. *255*, 6, 1797 (1980)
121. Emptage, M.M., et al.: ibid. *255*, 6, 1793 (1980)
122. Cammack, R.: Nature *286*, 442 (1980)
123. Ke, B., Hansen, R.E., Beinert, H.: Proc. Natl. Acad. Sci. *70*, 2941 (1973)
124. Evans, M.C.W., Reeves, S.G., Cammack, R.: F.E.B.S. Lett. *49*, 111 (1974)
125. Rupp, H., De la Torre, A., Hall, D.O.: Biochim. Biophys. Acta *548*, 552 (1979)
126. Evans, E.H. et al.: Biochim. J. *182*, 861 (1979)
127. Cammack, R., et al.: Biochim. Biophys. Acta *548*, 267 (1979)
128. Cheniae, G.M.: Ann. Rev. Plant Phys. *21*, 467 (1970)
129. Heath, R.L.: Int. Rev. Cyt. *34*, 49 (1973)
130. Trebst, A.: Ann. Rev. Plant Phys. *25*, 423 (1974)
131. Radmer, R., Kok, B.: Ann. Rev. Biochem. *44*, 409 (1975)
132. Radmer, R., Cheniae, G.M.: see Ref. 74, p. 303
133. Diner, B.A., Joliot, P.: see Ref. 7, p. 187
134. Joliot, P., Joliot, A.: Biochim. Biophys. Acta *153*, 625 (1967)
135. Joliot, P., Barbieri, G., Chabaud, R.: Photochem. Photobiol. *10*, 309 (1969)
136. Kok, B., Forbush, B., McGloin, M.: ibid. *11*, 457 (1970)
137. Mar, T., Govindjee: J. Theor. Biol. *36*, 427 (1972)
138. Kok, B., Velthuys, B.: see Ref. 2, p. 111
139. Inoue, Y., Shibata, K.: F.E.B.S. Lett. *85*, 2, 193 (1978)
140. Renger, G.: ibid. *81*, 2, 223 (1977)
141. Forbush, B., Kok, B., McGloin, M.: Photochem. Photobiol. *14*, 307 (1971)
142. Lavorel, J.: F.E.B.S. Lett. *66*, 2, 164 (1976)
143. Lavorel, J.: J. Theor. Biol. *57*, 171 (1976)
144. Thibault, P.: ibid. *73*, 271 (1978)
145. Jursinic, P.: Arch. Biochem. Biophys. *196*, 2, 484 (1979)
146. Fowler, C., Kok, B.: Biochim. Biophys. Acta *357*, 299 (1974)
147. Pirson, A., Tichy, C., Wilhelmi, G.: Planta *40*, 199 (1952)
148. Hopkins, E.F.: Science *72*, 609 (1930)
149. McHargue, J.S.: J. Am. Chem. Soc. *44*, 1592 (1922)
150. Bishop, W.B.S.: Aust. J. Exptl. Biol. Med. Sci. *5*, 125 (1928)
151. Emerson, R., Lewis, C.M.: Am. J. Bot. *26*, 808 (1939)

152. Emerson, R., Lewis, C.M.: ibid. *28*, 789 (1941)
153. Kessler, E.: Arch. Biochem. Biophys. *59*, 527 (1955)
154. Kessler, E., Arthur, W., Brugger, J.E.: ibid. *71*, 326 (1957)
155. Kessler, E., et al.: Biochem. Biophys. Acta *43*, 134 (1960)
156. Eyster, C., et al.: Plant Physiology *33*, 4, 235 (1958)
157. Cheniae, G.M., Martin, I.F., in: Brookhaven Symposia in Biology 19: Energy conversion by the photosynthetic apparatus, p. 406, 1966
158. Pirson, A.: Ann. Rev. Plant Physiol. *6*, 71 (1955)
159. Cheniae, G.M., Kok, B., in: Current Topics in Bioenergetics I (Ed. Sanadi, R.), Academic Press, New York, p. 1, 1966
160. Blankenship, R.E., Sauer, K.: Biochim. Biophys. Acta *357*, 252 (1974)
161. Blankenship, R.E., Babcock, G.T., Sauer, K.: ibid. *387*, 165 (1975)
162. Cheniae, G.M., Martin, I.F.: ibid. *253*, 167 (1971)
163. Radmer, R., Cheniae, G.M.: ibid. *253*, 182 (1971)
164. Takahashi, M., Asada, K.: Plant Cell Physiol. *18*, 807 (1977)
165. Ochai, E. (Ed.): Bioinorganic Chemistry: An Introduction, Allyn and Bacon, p. 388, 1977
166. Tolmaroff, J.M., Ono, T.E., Culter, R.Q.: Proc. Natl. Acad. Sci. *77*, 5, 2777 (1980)
167. Harris, J.I., et al.: Eur. J. Biochem. *106*, 297 (1980)
168. Beauchamp, L.O., Fridovich, I.: Biochim. Biophys. Acta *317*, 50 (1978)
169. Lumsden, J., Hall, D.O.: Biochim. Biophys. Res. Commun. *58*, 35 (1974)
170. Arron, G.P., et al.: Biochem. Soc. Trans. *4*, 618 (1976)
171. Asada, K., et al., in: Iron and Copper Proteins (Eds. Yasunobu, K.T., Mower, H.F., Hayaishi, O.), Plenum Press, New York, p. 551, 1976
172. Giannopolitis, C.N., Ries, S.K.: Plant Phys. *59*, 309 (1977)
173. Asada, K., Kanematsu, S., Uchida, K.: Arch. Biochem. Biophys. *179*, 243 (1971)
174. Del-Rio, L.A., et al.: Planta *140*, 221 (1978)
175. Kaniuga, Z., Zabek, J., Michalski, W.P.: ibid. *145*, 145 (1979)
176. Huzisige, H., Isimoto, M., Inove, H., in: Comparative Biochemistry and Biophysics of Photosynthesis, Papers of the conference (Eds. Shibata, K., et al.) Univ. Tokyo Press, Tokyo, p. 170, 1968
177. Cheniae, G.M., Martin, I.F.: Biochim. Biophys. Acta *502*, 321 (1978)
178. Velthuys, B., Amesz, J.: ibid *245*, 373 (1975)
179. Velthuys, B.: ibid. *396*, 392 (1975)
180. Govindjee, Wydrzynski, T., Marks, S.B.: see Ref. 106, p. 305
181. Lawrence, G.D., Sawyer, D.T.: Co-ordn. Chem. Revs. *27*, 173 (1978)
182. Coleman, W.M., Taylor, L.T., ibid. *32* (1980)
183. Kelley, P.M., Izawa, S., Biochim. Biophys. Acta *502*, 198 (1978)
184. Gribova, Z.P., Tikhonov, A.N.: Biophysics *22*, 4, 676 (1977)
185. McGarvey, B.R.: J. Phys. Chem. *61*, 1232 (1958)
186. Gol'dfel'd, M.G., et al.: Biophysics *23*, 2, 267 (1978)
187. Gribova, Z.P.: ibid. *22*, 1, 62 (1977)
188. Homann, P.: Biochem. Biophys. Res. Commun. *33*, 229 (1968)
189. Lozier, R., Baginsky, M., Butler, W.L.: Photochem. Photobiol. *14*, 323 (1971)
190. Siderer, Y., et al.: Arch. Biochem. Biophys. *179*, 174 (1977)
191. Meirovitch, E., Luz, Z., Kalb (Gilboa), A.J.: J. Am. Chem. Soc. *96*, 24, 7538 (1974)
192. Meirovitch, E., Luz, Z., Kalb (Gilboa), A.J.: ibid. *96*, 24, 7542 (1974)
193. Chien, J.C.W., Westhead, E.W.: Biochemistry *10*, 3198 (1971)
194. Reed, G.H., Ray, Jr., W.J.: ibid. *10*, 3190 (1971)
195. Reed, G.H., Leigh, Jr., J.S., Pearson, J.E.: J. Chem. Phys. *55*, 7, 3311 (1971)
196. Reed, G.H., Scrutton, M.C.: J. Biol. Chem. *249*, 10, 6156 (1974)
197. Berliner, L.J., Wong, S.S.: Biochemistry *14*, 22, 4977 (1975)
198. Michaels, G., Milner, Y., Reed, G.H.: ibid. *14*, 14, 3213 (1975)
199. Basosi, R., Tuzzi, C., Valensin, G.: J. Phys. Chem. *79*, 16, 1725 (1975)
200. Buttlaire, D.H., Reed, G.H., Himes, R.: J. Biol. Chem. *250*, 1, 261 (1975)
201. Bajaj, S.P., Nowak, T., Castellino, F.J.: ibid. *251*, 20, 6294 (1976)

202. Villafranca, J.J., Ash, D.E., Wedler, F.C.: Biochemistry *15*, 3, 544 (1976)
203. McLaughlin, A.C., Leigh, Jr., J.S., Cohn, M.: J. Biol. Chem. *251*, 9, 2777 (1976)
204. Wilson, Jr., G.G., Cohn, M.: ibid. *252*, 6, 2004 (1977)
205. Bagshaw, C.R.: Biochemistry *16*, 1, 59 (1977)
206. Buttlaire, D.H., Cohn, M., Bridger, W.A.: J. Biol. Chem. *252*, 6, 1957 (1977)
207. Markham, G.D., et al.: ibid. *252*, 4, 1197 (1977)
208. Meirovitch, E., Poupko, R.: J. Phys. Chem. *82*, 17, 1920 (1978)
209. Balokrishnan, M.S., Villafranca, J.J.: Biochemistry, *18*, 8, 1546 (1979)
210. Armstrong, R.N., et al.: ibid. *18*, 7, 1230 (1979)
211. Pillai, R.P., Marshall, M., Villafranca, J.J.: Arch. Biochem. Biophys. *199*, 1, 21 (1980)
212. Buttlaire, D.H.: J. Biol. Chem. *255*, 5, 2164 (1980)
213. Treharne, R.W., et al.: Biochem. Biophys. Res. Commun. *3*, 1, 119 (1960)
214. Siderer, Y., Malkin, S.: F.E.B.S. Lett. *104*, 2, 335 (1979)
215. Wydrzynski, T., Sauer, K.: Biochim. Biophys. Acta *589*, 56 (1980)
216. Cavé, A., et al.: Biochemie *61*, 755 (1979)
217. Wydrzynski, T., et al.: Biochim. Biophys. Acta *408*, 349 (1975)
218. Wydrzynski, T., et al.: Proc. Natl. Acad. Sci. U.S.A. *73*, 4, 1196 (1976)
219. Wydrzynski, T.J., et al.: Biochemistry *17*, 11, 2155 (1978)
220. Gol'dfel'd, M.G., Vozuyshayova, L.V., Yushmanov, V.Y.: Biophysics *24*, 2, 273 (1979)
221. Izawa, S., Heath, R.L., Hind, G.: Biochim. Biophys. Acta *180*, 388 (1969)
222. Gol'dfel'd, M.G., Khailova, I.I., Vanin, A.F.: Biophysics *24*, 570 (1979)
223. Muallem, A., Izawa, S.: F.E.B.S. Lett. *115*, 1, 49 (1980)

Carbonic Anhydrase: An Insight into the Zinc Binding Site and into the Active Cavity Through Metal Substitution

Ivano Bertini, Claudio Luchinat and Andrea Scozzafava

Istituto di Chimica Generale ed Inorganica della Facoltà di Farmacia, Università di Firenze, Florence, Italy

Structure and Bonding 48

A. Introduction

Carbonic anhydrase (CA) is a rather popular metalloenzyme, easy to handle, rather stable under a wide range of experimental conditions, and investigated with a large variety of techniques.

Excellent review articles are available which devote their major attention to the native enzyme[1-6]. In the recent years a number of papers dealing with physical chemical measurements of metal substituted derivatives of CA have appeared. Information have been obtained for example through electronic spectroscopy of chromogenic ions, EPR of paramagnetic ions, NMR of nuclei nearby the paramagnetic centers or directly of the metal ions which have a nuclear magnetic moment like ^{113}Cd.

The presence of the paramagnetic center or of the magnetic nucleus allows us to monitor the chemical properties of the moiety constituted by the donor atoms and the metal ion 1) as they depend on pH; 2) as they change upon interaction with the inhibitors; 3) as they change along the catalytic pathway with natural or artificial substrates; 4) in comparison with the chemical properties of simple inorganic compounds. The aim of the present article is to survey the meaningful results obtained in this area as compared with the known properties of the native enzyme.

B. The Native Enzyme

The enzyme is widespread in tissues of animal and vegetable species as well as in bacteria[1-17]. Mammalian blood contains about 1–2g/l of carbonic anhydrase[18-20] and, since human and bovine blood are easily available, the large majority of the literature data is concerned with enzymes from the above sources. In mammalian red cells the enzyme is constituted by a monomeric polypeptide of molecular weight around 30,000 and contains a single zinc(II) ion per molecule which is essential for the catalytic activity[21].

Several isoenzymes have been so far identified within each species: three different isoenzymes are present in human erythrocytes, which are labelled HCAA, HCAB, and HCAC, with a relative abundance of 5, 83, and 12% respectively[22-26] and two major components (BCAA and BCAB) in bovine erythrocytes[27]. Man is among a few animal species in which the isoenzymes are differentiated by large variations in catalytic activity; as a matter of fact the low activity forms A and B seem to be a later product of evolution, while the C form more closely resembles those found in other mammals[28]. The primary structure is known for both the B and C isoenzymes (Fig. 1)[29-36]. Despite the several changes in the aminoacid components on passing from human B to human C or bovine isoenzymes[37-38], many oligomers in the polypeptide chain remain unaltered[39], so that the secondary and tertiary structure is almost the same in all isoenzymes[40]. Therefore we will still refer to carbonic anhydrase as a single chemical species, unless there is a need to stress the differences between the various isoenzymes.

In the last decade X-ray data began to appear on carbonic anhydrase, and the X-ray structure is now resolved for several isoenzymes at about 200 pm nominal resolution[40-48]. From these data the rugby ball shape of the molecule, the large region of rigid β

```
     1              5                    10                  15                  20                  25                  30
 Ac-Ala     Pro-Asp             Asp-Asp-Lys                 Gln     Ser     Leu-Try             Asn     Asn-Asn
     Ac-Ser-His-His-Trp-Gly-Tyr-Gly-Lys-His-Asn-Gly-Pro-Glu-His-Trp-His-Lys-Asp-Phe-Pro-Ile-Ala-Lys-Gly-Glu-Arg-Gln-Ser-Pro
                                                                                                Asn

                     35                  40                  45                  50                  55                  60
                Lys     Ser-Glu-Thr     His     Thr                 Ile                 Asn-Pro         Ala-Lys-Glu     Ile
    Val-Asp-Ile-Asp-Thr-His-Thr-Ala-Lys-Tyr-Asp-Pro-Ser-Leu-Lys-Pro-Leu-Ser-Val-Ser-Tyr-Asp-Gln-Ala-Thr-Ser-Leu-Arg-Ile-Leu-
        Asn             Lys-Ala-Val-Val-Gln         Ala                 Ala-Leu-Val     Gly-Glu             Arg     Met-Val
                                                                                                           (Glu)

                     65                  70                  75                  80                  85                  90
        Val         Ser     His     Asn     Glu     Asn-Asn     Arg-Ser                         Phe-Ser-Asp-Ser
    Asn-Asn-Gly-His-Ala-Phe-Asn-Val-Glu-Phe-Asp-Asp-Ser-Gln-Asp-Lys-Ala-Val-Leu-Lys-Gly-Gly-Pro-Leu-Asp-Gly-Thr-Tyr-Arg-Leu-
                    Ser                 Tyr                                         Asp             Thr

                     95                  100                 105                 110                 115                 120
    Phe                                 Thr-Asn-Glu-His                             Gly-Val         Ser                 Val-
    Ile-Gln-Phe-His-Phe-His-Trp-Gly-Ser-Leu-Asp-Gly-Gln-Gly-Ser-Glu-His-Thr-Val-Asp-Lys-Lys-Lys-Tyr-Ala-Ala-Glu-Leu-His-Leu-
    Val                                 Ser     Asx                                 Arg

                     125                 130                 135                 140                 145                 150
    Ala             Ser-Ala         Ser-Ser-Leu-Ala-Glu     Ala-Ser-Lys-Ala                     Ile     Val-Leu-Met
    Val-His-Trp-Asn-Thr-----Lys-Tyr-Gly-Asp-Phe-Gly-Lys-Ala-Val-Gln-Gln-Pro-Asp-Gly-Leu-Ala-Val-Leu-Gly-Ile-Phe-Leu-Lys-Val-
                                                    Thr     Ala                                 Val     Val

                     155                 160                 165                 170                 175                 180
        Glu     Asn     Lys                 Leu     Ala     Gln-Ala                         Arg     Pro
    Gly-Ser-Ala-Lys-Pro-Gly-Leu-Gln-Lys-Val-Val-Asp-Val-Leu-Asp-Ser-Ile-Lys-Thr-Lys-Gly-Lys-Ser-Ala-Asp-Phe-Thr-Asn-Phe-Asp-
        Asp     Asn     Ala                 Leu     Ala                                         Thr         Pro

                     185                 190                 195                 200                 205                 210
        Ser-Thr             Ser             Phe                                 His             Tyr     Ser
    Pro-Arg-Gly-Leu-Leu-Pro-Glu-Ser-Leu-Asp-Tyr-Trp-Thr-Tyr-Pro-Gly-Ser-Leu-Thr-Thr-Pro-Pro-Leu-Leu-Glu-Cys-Val-Thr-Trp-Ile
        Gly-Ser             Asn-Val                                                                     Ser

                     215                 220                 225                 230                 235                 240
    Ile-Cys         Ser                             Leu-Ala-Gln         Ser     Leu-Ser     Val         Asp-Asn-Ala-Val-Pro-
    Val-Leu-Lys-Glu-Pro-Ile-Ser-Val-Ser-Ser-Glu-Gln-Val-Leu-Lys-Phe-Arg-Lys-Leu-Asn-Phe-Asn-Gly-Glu-Gly-Glu-Pro-Glu-Glu-Leu-
                                            Gln     Met                 Thr                 Ala                     Leu

                     245                 250                 255                 260
        Gln-His     Asn         Thr                 Gly     Thr-Val-Arg            -OH       HCAB
    Met-Val-Asp-Asn-Trp-Arg-Pro-Ala-Gln-Pro-Leu-Lys-Asn-Arg-Gln-Ile-Lys-Ala-Ser-Phe-Lys-OH   HCAC
        Leu-Ala                                                 Val-Arg-Gly-Phe-Pro    -OH   BCAB
                                                                                            (BCAA)
```

Fig. 1. Comparative primary structures of human and bovine carbonic anhydrase isoenzymes. The residues are aligned to obtain the highest number of homologies; the conventional numbering is that of the human B isoenzyme structure whose strands run all across the central part of the molecule, involving more than one third of the aminoacids, and the almost central position of the metal ion, sitting at the bottom of a roughly conical cavity about 1.5 nm deep, (Fig. 2), were soon apparent.

I. Around the Active Site

For a better understanding of the properties of the enzyme including the possible stereochemistries of the metal ion, the catalytic mechanism and the interactions with inhibitors, a closer inspection of the metal ion environment is needed. High resolution X-ray data[40–48] show that the basis of the active site cavity is constituted by four strands of the β structure indicated by γ, δ, ε, and ζ (Fig. 2); His 94 and 96 from strand δ and His 119 from strand ε coordinate the zinc ion, the former two with the nitrogen in position 3 and the latter with the nitrogen in position 1. His 94 is a little bit further than expected for regular coordination, the metal lying also some 30° out of the imidazole ring plane. In Table 1 the metal-nitrogen distances and the bond angles are reported; if allowance is made for the large errors on the bond distances and presumably more on the angles, these data may appear to be consistent with a pseudo-tetrahedral structure. A fourth weak electron density is observed in the vicinity of the zinc atom; it is assigned to a water molecule or a hydroxide ion, since it is poorly connected with the protein backbone. The resultant coordination polyhedron is represented in Fig. 3.

Outside the first coordination sphere of the metal ion, the first important feature encountered is the OH group belonging to Thr 199, which is almost opposite to His 94

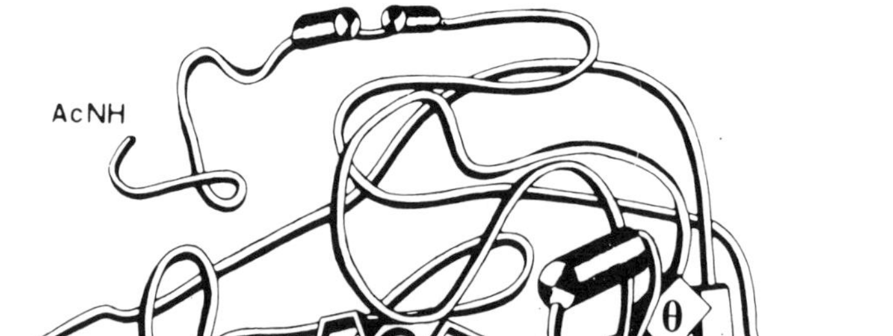

Fig. 2. Secondary and tertiary structure of human carbonic anhydrase C. The cylinders represent the α helices and the arrows the strands of the β structure. The zinc atom together with its histidine ligands is also shown

Table 1. Metal nitrogen distances and bond angles in human carbonic anhydrase C[a]

	R (pm)	
N (His 94)	240	$N_{94} - \hat{Zn} - N_{96} = 117^0$
N (His 96)	200	$N_{94} - \hat{Zn} - N_{119} = 113^0$
N (His 119)	200	$N_{96} - \hat{Zn} - N_{119} = 121^0$

[a] Based on Ref. 48

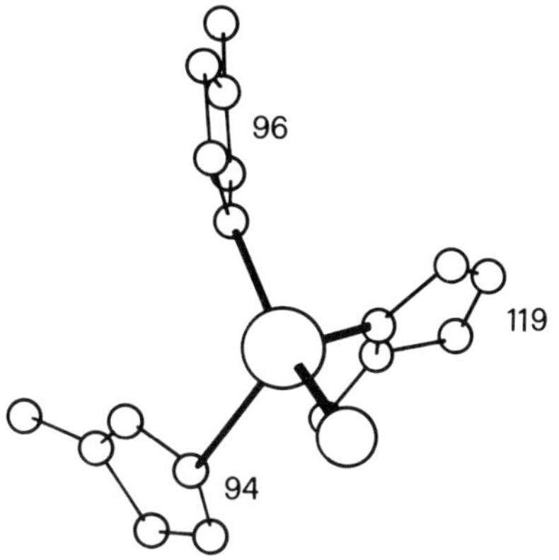

Fig. 3. Coordination polyhedron of the zinc atom in human carbonic anhydrase C showing the reciprocal orientation of the three histidyl ligands and the water molecule

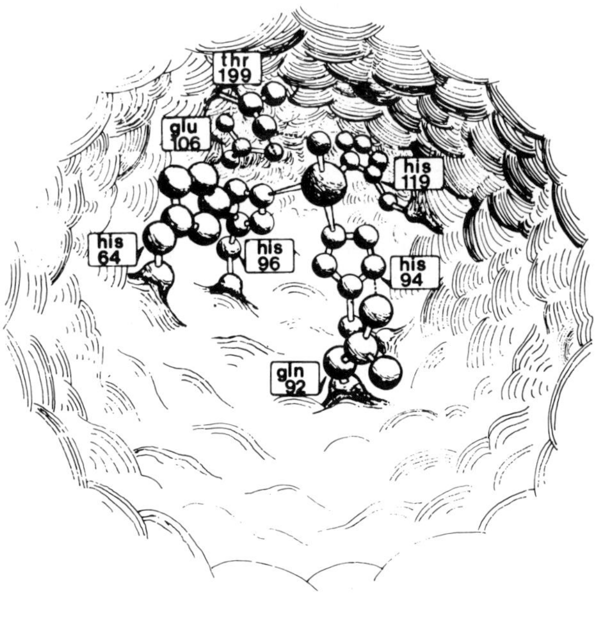

Fig. 4. Pictorial representation of the active site cavity showing the residues in close proximity of the zinc atom which are common to all the human and bovine isoenzymes

and within hydrogen bonding distance to both the metal coordinated water molecule and the carboxyl group of Glu 106, the latter being part of the fragment connecting δ and ε strands and almost buried behind the bottom of the cavity. By further moving towards the outside of the active site other residues of interest are encountered, namely His 64 and Gln 92. Figure 4 is a pictorial representation of the inner part of the cavity, showing the reciprocal orientation of all the above mentioned residues. All of them are present in all the carbonic anhydrase isoenzymes. The overall shape of the cavity is slightly different in human B and C isoenzymes, owing to the presence of different residues in that area: the most significant differences are at position 67 and 200 (Fig. 1). In particular in the low activity human B isoenzyme two more histidines are present in the place of Asn 67 and Thr 200. This reduces the volume of the cavity available to the solvent and to solute molecules. The bovine isoenzymes are strictly similar to the human isoenzyme C[49)].

Several chemical modifications have been successfully performed on carbonic anhydrase[50–58)]. The human B form can be carboxymethylated at His 200[50–57)] while human carbonic anhydrase C has been carboxyketoethylated at His 64[58)].

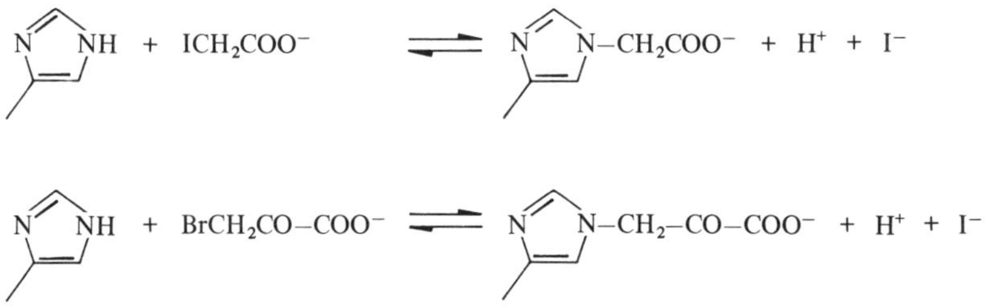

All of these manipulations cause drastic reduction but not complete removal of catalytic activity[56–58)], suggesting that none of the above residues are essential for the catalytic process. This is not unexpected for His 200 which is present only in human carbonic anhydrase B, since it is reasonable to assume a common catalytic pathway for all the isoenzymes and therefore common residues to be involved in the catalytic steps should be looked for.

II. Catalysis

Since the first description by Meldrum and Roughton in 1932[61)] of an enzyme in blood able to catalyze the reversible hydration of CO_2, which they named carbonic anhydrase, no other physiological function has been discovered for this enzyme. Only in the sixties other activities[62–90)] have been found "in vitro", such as hydrolysis of esters[62–78)] and sultones[82–85)], and hydration of aldehydes[86–90)].

$$CO_2 + H_2O \rightleftharpoons HCO_3^- + H^+ \quad (1)$$

$$R\text{–}COOR' + H_2O \rightleftharpoons RCOO^- + R'OH + H^+ \quad (2)$$

$$RSO_3R' + H_2O \rightleftharpoons RSO_3^- + R'OH + H^+ \quad (3)$$

$$RCHO + H_2O \rightleftharpoons RCH(OH)_2 \quad (4)$$

A large amount of experimental work has been done through different kinetic techniques to elucidate the effect of the enzyme on reaction (1), although discrepancies have been registered among the published data.

The first, uncontroverted point to be stressed is the extremely high efficiency of the enzyme. The uncatalyzed first order rate constant for the hydration reaction is around $3 \times 10^{-2}\ s^{-1}$, although above pH 10 the reaction $CO_2 + OH^- \rightleftharpoons HCO_3^-$ predominates with a second order rate constant of $8.5 \times 10^3\ s^{-1}$ [91–97]: at neutral pH the enzyme is able to enhance the hydration rate by a factor of 10^7 [98]. The second experimental evidence is a drastic pH dependence of the catalytic efficiency: the hydration activity increases and the dehydration activity decreases with increasing pH[99–103]. The pH rate profiles for both functions are roughly of sygmoidal type at least in buffered solutions and in a limited pH range; they have been taken as indicative of a single acid-base equilibrium in the active site whose pK_a, however, ranges from 6.1 to 8.4 depending on the experimental conditions and the particular isoenzyme used[99–104]. The overall catalytic rate at a given enzyme concentration depends on two factors:

1) the Michaelis constant, K_M, which determines how many substrate molecules at fixed enzyme concentration are undergoing the catalytic process.
2) The rate of the catalytic process when the enzyme is fully saturated with the substrate, which is called V_{max}.

In the case of the hydration reaction K_M has been shown to be independent of pH whereas V_{max} is not[101–109]; the observed pH dependence of the catalytic rate is therefore attributable to the latter parameter. In the case of the dehydration reaction conflicting reports have been published[99, 104, 107]; however at least in the case of the bovine enzyme it has been shown that the only pH dependent parameter is again V_{max}[101, 109].

The above data have lead to the common statement that the "acidic form" of the enzyme is required for the dehydration, and the "basic form" for the hydration reaction; this is incorrect and violates the thermodynamic definition of the enzyme[110], unless it is stressed that the reaction

$$CO_2 + H_2O \rightleftharpoons HCO_3^- + H^+$$

involves a proton, which can be bound to the enzyme at some step of the catalytic pathway

$$CO_2 + H_2O + E \rightleftharpoons HCO_3^- + EH^+ \rightleftharpoons HCO_3^- + E + H^+$$

where EH^+ and E represent the acid and basic forms of the enzyme.

While this simple view held for a long time, and has been undoubtedly useful in rationalizing many of the enzyme chemical properties, several more subtle effects[50, 102, 111, 112] remained unexplained and lead to more complex theories. Recently thorough studies of isotope effects on the kinetic parameters have lead to the postulation of the existence of a second ionizing group[107, 113–117] as well as the occurrence of an internal proton transfer during the catalytic pathway; the role of buffers in accelerating the interconversion between EH^+ and E is also becoming apparent[93, 117–125]: this interconversion becomes the rate limiting step for buffer concentrations lower than 10^{-3} M[125].

The other hydrolytic activities of carbonic anhydrase seem to follow the same pH dependence as the CO_2 hydration[66, 71, 73, 74, 87, 126]; since the ester hydrolysis is much easier to follow experimentally, the hydrolysis of p-NO_2 phenylacetate has become a widespread assay method for carbonic anhydrase[63, 65, 77]. Although not definitely proved, the reactions 1 through 4 are likely to involve the same catalytic groups. It should be noted that the aldehyde hydration reaction does not require the release of a proton. The observed increase of the activity at high pH[87, 126] is thus related to the increase in concentration of the alkaline form of the enzyme; the latter must be therefore considered the only active enzyme species. Unfortunately, there is no way to check this hypothesis by studying the dehydration reaction in this case.

III. Inhibition

Carbonic anhydrase is reversibly inhibited by a large number of simple anions, which are summarized in Table 2. All of them are non competitive with respect to the hydration reaction[91–97] i.e. they bind the protein at a different site from that of CO_2. The apparent affinity constants of anions for the enzyme as a whole decrease with increasing pH[100]. Several authors[68, 69, 144] stated that the pH dependence of the affinity constants of the anions is sygmoidal with a pK_a around 7 like that of the group controlling the catalytical activity and proposed the following scheme to account for the observed pH dependence.

$$\begin{array}{ccc} & EH^+ \underset{}{\overset{K_a}{\rightleftharpoons}} E & \\ K_I & \Big\Updownarrow \quad +I^- & \\ & EH\text{–}I & \end{array} \quad ; \qquad K_{app} = K_I \frac{1}{1 + \dfrac{K_a}{[H^+]}} \qquad (1)$$

Recently[145] accurate investigations on the cobalt(II) substituted CA pointed out that the affinity of the anions in absence of buffers becomes larger and larger as the pH is lowered, the resulting dependence of K_{app} being no more represented by a sygmoid curve. However, if the range of pH investigated is restricted, and if the experiments are performed in the presence of interacting buffering species, as were the cases referred by the previous authors, the apparent affinity constants follow with reasonable accuracy a sygmoidal dependence on pH. When the concentration of the anion is pH dependent because it is the conjugated base of an acid with a pK_a falling in the pH range of interest (6–10) the profile of K_{app} against pH is bell shaped[69]; for example the apparent affinity for the enzyme of cyanide (as a substance), will decrease at low pH because the concentration of the CN^- ion decreases according to the value of pK_a of HCN = 9.3 (Fig. 5).

In such cases the maximum of the bell shaped profile of K_{app} with pH falls at pH = 1/2 (pK_{a1} + pK_{a2}) were 1 and 2 refer to the enzyme and to the inhibiting species respectively[69]. K_{app} can be expressed as

$$K_{app} = K_I \frac{1}{1 + \dfrac{K_{a1}}{[H^+]}} \cdot \frac{1}{1 + \dfrac{[H^+]}{K_{a2}}} \qquad (2)$$

$$EH^+ \overset{K_{a1}}{\rightleftharpoons} E + H^+$$

$$HI \overset{K_{a2}}{\rightleftharpoons} I^- + H^+$$

$$EH^+ + I^- \overset{K_I}{\rightleftharpoons} EH\text{–}I$$

Table 2. Affinity constants of some inhibitors of the native enzymes for several metallocarbonic anhydrases[a]

	BCA	HCAC	HCAB	CoCA	CuBCA	MnBCA	NiBCA	CdHCAB
HS^-	5.72[69) b] 4.95[127) b]			5.80[135) k]				
CN^-	5.49[69) b] 5.58[127) b]			> 5[136) i, l]	> 8[140) l, m]			
NCO^-	4.41[69) b] 3.96[127) b]			5.15[137) l, m]	5.48[140) l, m]		> 3[143) l]	
NCS^-	4.04[127) b]	3.09[68) b, e]	3.15[68) b, j]	3.80[136) l, n]	5.65[140) l, m]		> 3[143) l]	
N_3^-	3.23[127) b]			3.60[136) l]	6.00[140) l, m]	3.50[141) r]	> 3[143) l]	
I^-	2.06[127) b]	2.51[68) b, e]	2.70[68) b, j]	3.04[137) m, l]	4.27[140) l, m]		2.49[143) l, t]	
ClO_4^-	1.80[127) b]	2.82[68) b, e]	2.89[68) b, j]					
HCO_3^-	1.58[127) b]			1.00[136) k, l]	3.54[140) l, m]			
HSO_3^-	1.52[127) b]			> 5[136) f, l]				
NO_3^-	1.32[127) b]	1.74[68) b, e]	1.82[68) b, j]	3.52[137) l, m]				
Br^-	1.18[127) b]	1.57[68) b, e]	1.64[68) b, j]	2.10[137) l, m]	2.85[140) l, m]		2.10[143) l, u]	
CH_3COO^-	1.07[127) b]	1.47[68) b, e]	1.58[68) b, j]	2.10[136) l]		−0.08[142) b] 1.50[142) r, s]		
Cl^-	0.72[127) b]	1.17[68) b, e] 0.70[130) b, f]	1.30[68) b, j] 2.70[116) b, f]	1.70[137) l, m]	2.23[140) l, m]			3.55[139) b, v]
F^-	−0.08[127) b]		0.40[68) b, j]	1.60[137) l, m]				
$(COO)_2^{2-}$				2.50[136) l]			> 3[143) l]	
$HCOO^-$		1.74[131) g]		2.68[131) g, o]				
CH_2FCOO^-		0.89[131) g]		1.92[131) g, o]				
CHF_2COO^-		1.30[131) g]		2.15[131) g, o]				
CF_3COO^-		1.96[131) g]		1.96[131) g, o]				
Acetazolamide	6.70[127) b] 7.65[128) c, d]	7.10[132) h]	7.10[132) h]	6.82[132) h, p]	5.15[132) h, p]	5.60[132) h, p]	4.30[132) h, p]	4.80[132) h, p]
p-toluenesulfonamide	6.32[129, b]	7.09[133, i]				4.44[141) r]	> 3[143) l]	
Sulfanilamide	5.35[129) b] 4.96[127) b]	4.55[134) c]	5.49[134) c]	5.34[138) b, q]				
Phenol	1.52[127) b]							
Pyridine	0.3[127) b]							
Aniline	1.55[127) b]			1.17[136) l]				
Imidazole			1.70[102) c]	1.30[139) l, p]				1.52[139) w]

[a] Apparent affinity constants (reported as log K_{app}) at pH 7.5 if not otherwise stated; [b] Data obtained from the inhibition of the p-nitrophenyl acetate hydrolysis; [c] Data obtained from the inhibition of carbon dioxide hydration; [d] pH = 6.7; [e] pH = 6.8; [f] pH = 6.5; [g] pH = 7.6. Data obtained through fluorescence affinity titrations; [h] pH = 8.0 using the equilibrium dialysis technique; [i] pH = 8.0; [j] pH = 7.3; [k] pH = 8.2; [l] Values obtained from spectrophotometric measurements; [m] Actual affinity constant (log K_I); [n] pH = 8.1; [o] Human C isoenzyme; [p] Human B isoenzyme; [q] pH = 7.0; [r] Data obtained through nmr measurements of linewidth broadening; [s] The discrepance with the value obtained through inhibition measurements has been attributed to the presence of a non inhibitory second site. (Ref. 142); [t] pH = 6.4; [u] pH = 6.1; [v] pH = 9.5; [w] pH = 8.5 using sultone as substrate

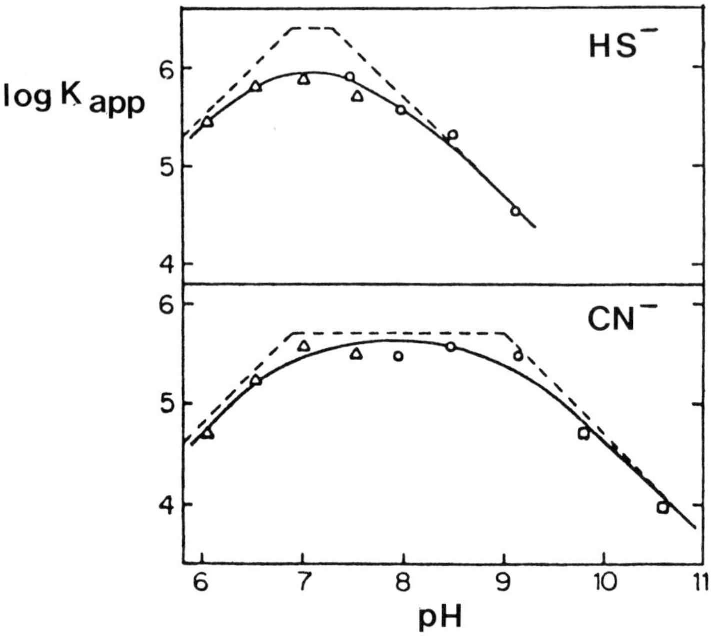

Fig. 5. pH dependence of the apparent affinity constants of hydrogen sulfide and cyanide for bovine carbonic anhydrase

Is should be noted that in principle the same pH dependence of K_{app} is expected if the alkaline form of the enzyme (E) reacted with the neutral form of the inhibitor (HI)

$$E + HI \overset{K_I}{\rightleftharpoons} E\text{–}HI$$

Aromatic sulfonamides constitute another important class of carbonic anhydrase inhibitors, generally showing high affinity towards the enzyme (see Table 2)[146–150]; their inhibition is also of non competitive type and the pH dependence of their apparent affinity constants is bell shaped like that of cyanide[151]. Although no conclusive evidence has been reached on the nature of the sulfonamide-enzyme adduct[152, 153] many experimental data suggest that sulfonamides bind the metal as anions[133, 149–151, 153–158]. The existence of strong hydrophobic interactions between the aromatic moiety of sulfonamide molecules and the protein part of the enzyme has been also shown[41, 129, 155]. Kinetic studies have shown that the association rate displays the same pH dependence as the thermodynamic constant[151]. On the assumption that the reaction is a first order process the thermodynamic constant is given by the ratio betweeen the association and dissociation rate constants. This implies that the pH dependence of the affinity is governed by the association rate.

It should be noted that sulfonamides, as well as some of the anions reported in Table 2, like NO_3^- or ClO_4^-, are very poor coordinating agents for metal ions in aqueous solution[159, 160]. However, there is no doubt that inhibitors do bind the metal ion in carbonic anhydrase (vide infra). In particular infrared studies on the N_3^- adduct of BCA[161] showed that the asymmetric stretching mode of the bound anion was strongly shifted with respect to the free anion absorption, indicating a direct binding to the metal. ^{13}C NMR investigations of cyanide ion in the presence of HCAB and BCA[162] demonstrated that the carbon atom of this ion is directly bound to the zinc. Several ^{35}Cl[163–165] and ^{81}Br[165, 166] nuclear relaxation enhancement studies indicated again a direct interaction of these ions with the zinc atom. Finally, EXAFS studies allowed to determine a zinc-iodide distance of 265 pm in the lyophylized iodide-BCAB adduct[167]. A further evidence of direct interaction between the inhibitors and the zinc(II) ion comes from the observation that removal of the zinc from carbonic anhydrase drastically reduces the affinity of inhibitors[4].

Neutral molecules other than sulfonamides generally show little affinity for the enzyme, and somewhat puzzling behavior. Aniline apparently binds the acidic form of the enzyme[127, 168–170] while methanol binds the basic form, although it is possible that the latter does not bind the metal[170, 171]. Imidazole is not an inhibitor of the high activity bovine and human C isoenzymes, while it inhibits the low activity human B form[102, 105, 139, 172]. Furthermore, imidazole has been reported to show competitive behavior and, apparently, do bind both the low and high pH forms[102, 105, 139].

Up to now no mention has been made of the nature of the ionizing group or groups controlling the pH dependence of inhibition and catalytic activity. While this point will be discussed in the following sections, we simply list here the possible candidates:

1. the zinc bound water molecule[93, 173] whose presence is inferred from the X-ray data[42–48]
2. one of the coordinated histidines which would ionize to imidazolate anion[174–177]
3. a non coordinated histidine (e.g. His 64) whose pK_a is around 7[178]
4. a glutammic residue (namely Glu 106) which could loose a proton through the hydrogen bonding net which links it to Thr 199 and to the zinc bound water molecule[179].

C. Metal Substitution

The zinc(II) ion can be removed from the enzyme by lowering the pH down to 5 and by addition of zinc chelating agents, to yield an almost inactive apoenzyme[23, 135, 180–183]. Extensive studies on the kinetics and thermodynamics of the process

$$\text{apoenzyme} + Zn^{2+} \rightleftharpoons \text{native enzyme}$$

as well as its pH dependence have been carried out[180, 184–187]; the results are summarized in Table 3. The data are qualitatively consistent with a tridentate behavior of the protein ligand with respect to Zn^{2+}.

The apoenzyme is able to bind a number of different metal ions, such as VO^{2+}, Mn^{2+}, Fe^{2+}, Co^{2+}, Co^{3+}, Ni^{2+}, Cu^{2+}, Cd^{2+} and Hg^{2+} in a 1:1 ratio[180, 188, 191, 192] (Table 3). Difference electron density maps have shown that Co^{2+}, Cu^{2+}, Mn^{2+}, are bound essentially in the same place as the native zinc(II) ion[40, 48, 193]. The mercury atom in the dichloromercury(II) derivative is displaced by about 60 pm from the Zn(II) position[193]. Besides the above reported metal ions, Fe^{3+}, Pb^{2+}, Be^{2+}, Mg^{2+}, Ca^{2+} and Ba^{2+} have also been tested[180, 194] for their ability to reconstitute a catalytically active enzyme. Only the

Table 3. Stability constants and kinetical data for some metallocarbonic anhydrases

	Log K (M^{-1})	k_d (s^{-1})[a]	k_f ($M^{-1}s^{-1}$)[b]
ZnBCA	9.9[188) c] 12.0[188) d]	1×10^{-6} [183) c]	1.5×10^{4} [186) d]
ZnHCAB[180) e]	10.5		
CoBCA[189) c]	6.0	1.2×10^{-6}	7.2×10^{-1}
CoHCAB[180) e]	7.2		
MnBCA[190)]	3.3[f]	1.1×10^{-3} [g] 3.5×10^{-4} [h]	9.0[g] 56[h]
MnHCAB[180) e]	3.8		
VOBCA[191)]	11.4[i]	6.8×10^{-4} [l]	
CuHCAB[180) e]	11.6		
NiHCAB[180) e]	9.5		
CdHCAB[180) e]	9.2		
HgHCAB[180) e]	21.5		

[a] Rate constant for the dissociation of the metallocarbonic anhydrase; [b] Rate constant for the formation of the metallocarbonic anhydrase; [c] at pH 5.0; [d] at pH 7.0; [e] at pH 5.5; [f] at pH 5.9; [g] at pH 6.25; [h] at pH 7.75; [i] at pH 8.5; [l] at pH 7.9

Table 4. Relative enzymatic activities of some human metallocarbonic anhydrases[a]

	Hydration of CO_2	Hydration of p-NO_2 phenylacetate[b]
ZnHCAB	100	100
ZnHCAC	294	270
CoHCAB	56	322
CuHCAB	1	18
NiHCAB	5	12
MnHCAB	4	14
CdHCAB	4	7
HgHCAB	0	3

[a] Based on Ref. 132; [b] At pH 7.5. At alkaline pH the cadmium derivative displays 30% of the ZnHCAB activity[195)]

cobalt(II) ion is able to give rise to a derivative with activity comparable to that of the native enzyme, the other metalloderivatives being less than 10% active (Table 4). The catalytic activities towards CO_2 hydration and ester hydrolysis do not strictly parallel each other[132, 194)]. A possible source of error in determining the catalytic efficiency of low activity derivatives can be the presence of impurities of native zinc enzyme; it follows that some of the smallest numbers of Table 4 should be taken as upper limit values. The catalytic activity of the cobalt(II) derivative shows a roughly sygmoidal pH dependence similar to that of the native zinc enzyme, the pK_a being also quite similar[137)]. A sygmoidal behavior is found for some of the less active derivatives but, generally, higher pH values are required with respect to zinc and cobalt enzymes to obtain measurable effects[195, 196)]. Inhibitors of the native CA often act as ligands of the metallosubstituted derivatives; they still display their inhibiting properties of the residual activity.

D. Water Proton Relaxation

In order to propose any reasonable mechanism for the catalytic action of CA it is of primary importance to check the presence and the number of water molecules in the coordination sphere of the metal in enzyme solutions.

Since most of the enzyme properties have been found to be strongly pH dependent, it would be desirable to ascertain independently the presence of solvent molecules bound to the metal ion all over the pH range of interest, their possible acid/base properties, and their interaction with metal binding inhibitors. Substitution of the native zinc(II) ion with paramagnetic 3d metal ions among those listed in the preceeding section allows NMR techniques to be applied to the elucidation of the problem: the approach is essentially based on the strong coupling between water protons and unpaired electrons of a paramagnetic metal ion, which causes sizeable shortening of the ligand nuclear relaxation times.

In order to detect the presence of water in the coordination sphere of the metal in the enzyme by measuring the water proton relaxation times two main conditions must be fulfilled[197–199]:

1. The exchange time of the water protons from the paramagnetic site to the bulk solution is smaller than the relaxation times they experience in the metal coordination sphere;
2. The relaxing capability of the metal ion is high enough to give measurable effects when averaged over the total number of water protons in solution.

Early data on cobalt(II) substituted bovine and human carbonic anhydrases did reveal the presence of exchangeable protons in the metal coordination sphere at alkaline pH[200], in agreement with the X-ray structure of the crystalline native enzyme prepared at pH 8.7; however, the paramagnetic effect decreased by decreasing pH, leading to the suggestion that in the acidic form of the enzyme water was not in fast exchange with the bulk solution, or not coordinated at all. This result raised a major debate, which is still alive, on the validity of the simplest model for the active site ionization, i.e. the deprotonation of the coordinated water molecule.

Analogous results were later obtained on the manganese(II) derivative[196, 201]: the paramagnetic effect was only present at high pH, and was again considered to be consistent with a metal bound water molecule rather than with a hydroxide ion.

Thorough studies of the water proton relaxation in the presence of bovine carbonic anhydrase substituted with a number of paramagnetic metal ions[136, 143, 201–210] and at various magnetic fields[196, 200, 211, 212], have allowed to obtain a more complete picture of the behavior of water in the metal coordination sphere upon pH variations and inhibitor binding. The longitudinal relaxation rates, T_1^{-1}, for water protons in the presence of several metallosubstituted carbonic anhydrases as a function of pH are reported in Fig. 6: it appears that exchangeable protons are present in the coordination sphere of all the metal derivatives investigated. The large difference in the observed $^1H\ T_1^{-1}$ values for the various metal derivatives is not unexpected, since the relaxing capability is strictly related to the metal electronic relaxation times, as will be discussed later.

At variance with the earlier results, no major pH dependence is detected in any of the derivatives investigated, provided that the enzyme solution is free from anions. As a matter of fact the earlier data[200] on cobalt carbonic anhydrase were altered by the

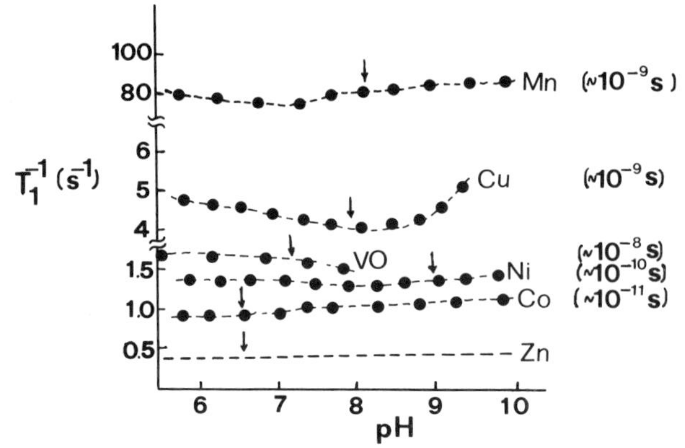

Fig. 6. ^{1}H relaxation rates of water solutions containing 1×10^{-3} M metallocarbonic anhydrases as a function of pH. The *arrows* indicate the pK_a of the metal linked acid-base equilibrium. In parentheses the common values of the metal electronic relaxation times are reported

presence of sulfate[136, 203, 207, 210], which binds the metal ion at low pH as such or as HSO_4^- [210] and probably removes the water molecule. The decrease of the paramagnetic effect at low pH previously observed in the manganese derivative[196, 201] was probably due to dissociation of Mn^{2+} ions from the protein, consistent with the relatively low stability of the complex; if more concentrated solutions are used, and the contribution of the above equilibrium is taken into account, exchangeable protons can be detected all over the pH range 6–10[204].

If the T_1^{-1} values reported in Fig. 6 are diminished by the relaxation rate measured in solutions containing equal amounts of diamagnetic zinc enzyme, the quantity obtained, T_{1p}^{-1}, represents the neat contribution of the paramagnetic center to the proton relaxation rate, which can be expressed by the Solomon, Bloembergen and Morgan (SBM) equation[213–216]:

$$\frac{f}{n} T_{1p}^{-1} = T_{1M}^{-1} = \frac{K}{r^6} f(\tau_c) + K' f(\tau_e)$$

where f is the molar ratio between the water protons and the enzyme and n is the number of protons interacting with the paramagnetic center at distance r; K and K′ are products of physical constants; τ_e is the electronic relaxation time of the metal ion and τ_c is a correlation time which is related to τ_e and to the rotational time of the complex, τ_r, through the relation; $\tau_c^{-1} = \tau_e^{-1} + \tau_r^{-1}$. The rotational time of a macromolecule of the size of carbonic anhydrase is of the order of 10^{-8}s[129], while the electronic correlation times τ_e range between 10^{-8} and 10^{-13}s for the metal ions investigated[197–217]. Therefore τ_c is determined by the latter term. From the above equation it would be possible, in principle, to evaluate the number of protons n in the coordination sphere of the metal; unfortunately, there are too many parameters in the equation which are only approximately known to be confident of the numerical result, although there are attempts in this sense reported in the literature[199]. Presently, it can only be stated that the presence of two protons, i.e. of a water molecule, can be consistent with the obtained data. The $f(\tau_c)$ part of the SBM equation is reported in Fig. 7 as a function of τ_c. The dots represent the values obtained from the experimental data on the different metal derivatives on the assumption that a single water molecule is bound with metal-hydrogen distance of

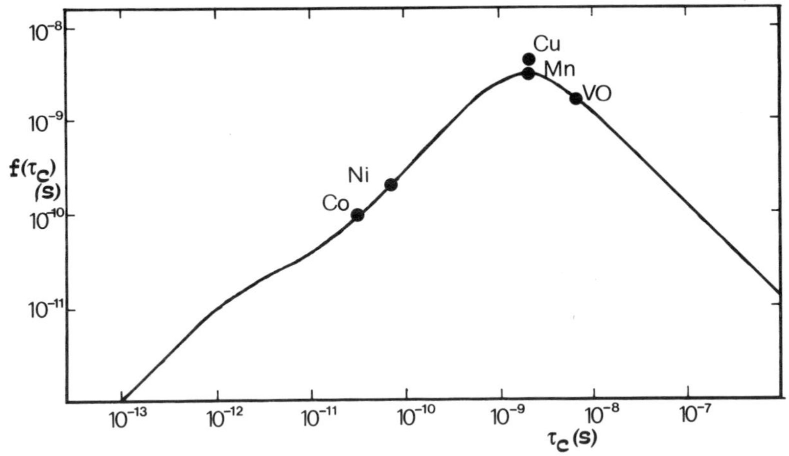

Fig. 7. $f(\tau_c)$ part of the SBM equation as a function of τ_c. The *dots* represent the $f(\tau_c)$ values obtained from the relaxation data on the various metal derivatives by assuming r = 280 pm and n = 2 (n = 4 in the case of NiBCAB)

280 pm (in the nickel(II) derivative two water molecules are assumed to be bound, vide infra). The τ_c values obtained are consistent with the usual values of the electronic relaxation times of the above metal ions, and thus in qualitative agreement with the above hypothesis.

The pH independence of the paramagnetic effect is apparently in favour of the presence of a water molecule all over the pH range, since a decrease at high pH would be expected upon ionization to a hydroxide ion, if the other parameters in the equation remain the same. However, a 14% decrease in metal-proton distance in the hydroxo complex would be enough to counterbalance the loss of a proton in the total relaxation effect. On the other hand, it is known that coordinated hydroxide ions display much longer exchange times as compared to water molecules[218–220] and therefore a metal bound hydroxide would not contribute to the water proton relaxation, unless hydrogen bonding of the hydroxo group with a second sphere water molecule provides a pathway for hydrogen exchange; such a mechnism has been proposed to occur in simple inorganic complexes[220, 221]. Finally, there is the possibility of addition of a hydroxide ion at high pH without displacement of the water molecule, increasing the metal coordination number from four to five; an increase in the number of protons from two to three could easily be masked by variations in r and/or in the electronic relaxation times of the resulting chromophores. Relaxation rates of ^{17}O would in principle be capable of sheding light on the problem. However, no significant differences in water ^{17}O relaxation rates were observed between zinc and cobalt substituted carbonic anhydrase[222] indicating that owing to the large water to metal ratios the neat contribution of the paramagnetic center is small as compared to other relaxation mechanisms experienced by ^{17}O in protein solutions (see Note Added in Proofs 1).

The water proton relaxation could also be used to ascertain the presence of water in the metal coordination sphere of the inhibitor derivatives of some metallo carbonic anhydrases allowing definition, in some cases, of the stereochemistries and coordination numbers (see following sections).

E. The Coordination Around the Metal Ions

The knowledge of coordination numbers and stereochemistries of various metal derivatives may be relevant with respect to the donor properties and flexibility of the active cavity. The relevance is particularly evident in the case of the active cobalt derivatives whose results can be transferred to the native enzyme with some confidence. A comparison of the stereochemical differences among the various metallo substituted derivatives as related to their catalytic efficiency may also be meaningful. A further point of interest is the stereochemistry of the adducts of the various metallosubstituted carbonic anhydrases with the molecules which act as inhibitors of the native enzyme. In general the comparison of the physical chemical properties of all the derivatives of each metallo substituted CA is by itself quite helpful in understanding the coordination chemistry of the compounds.

I. Stereochemistry of the Cobalt(II) Derivatives

The electronic absorption spectra of CoCA in the range $8 - 23 \times 10^3$ cm^{-1} are well shaped and strongly pH dependent (Fig. 8)[135, 136, 194, 223]. The profile of the absorption at 15,600 cm^{-1} against pH (Fig. 9) shows that more than one acid base equilibrium is capable to affect the electronic spectra, although the effect of one equilibrium is predominant over the others[224–226]. The pK_a of such equilibrium is located between pH 6 and pH 7.7 depending on the particular isoenzyme.

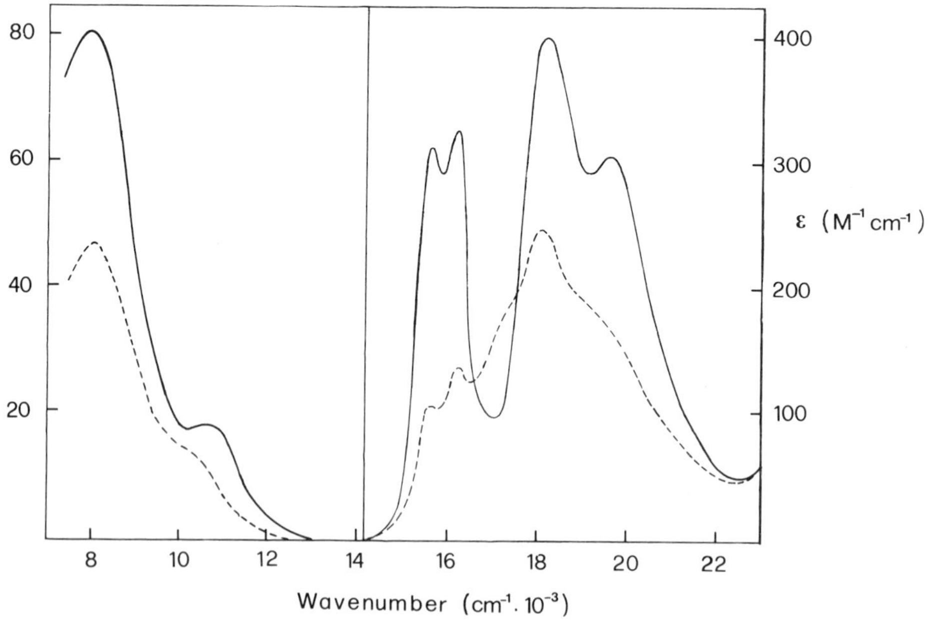

Fig. 8. Electronic absorption spectra of CoBCAB in unbuffered solutions at pH 5.8 (-----) and 8.8 (——)

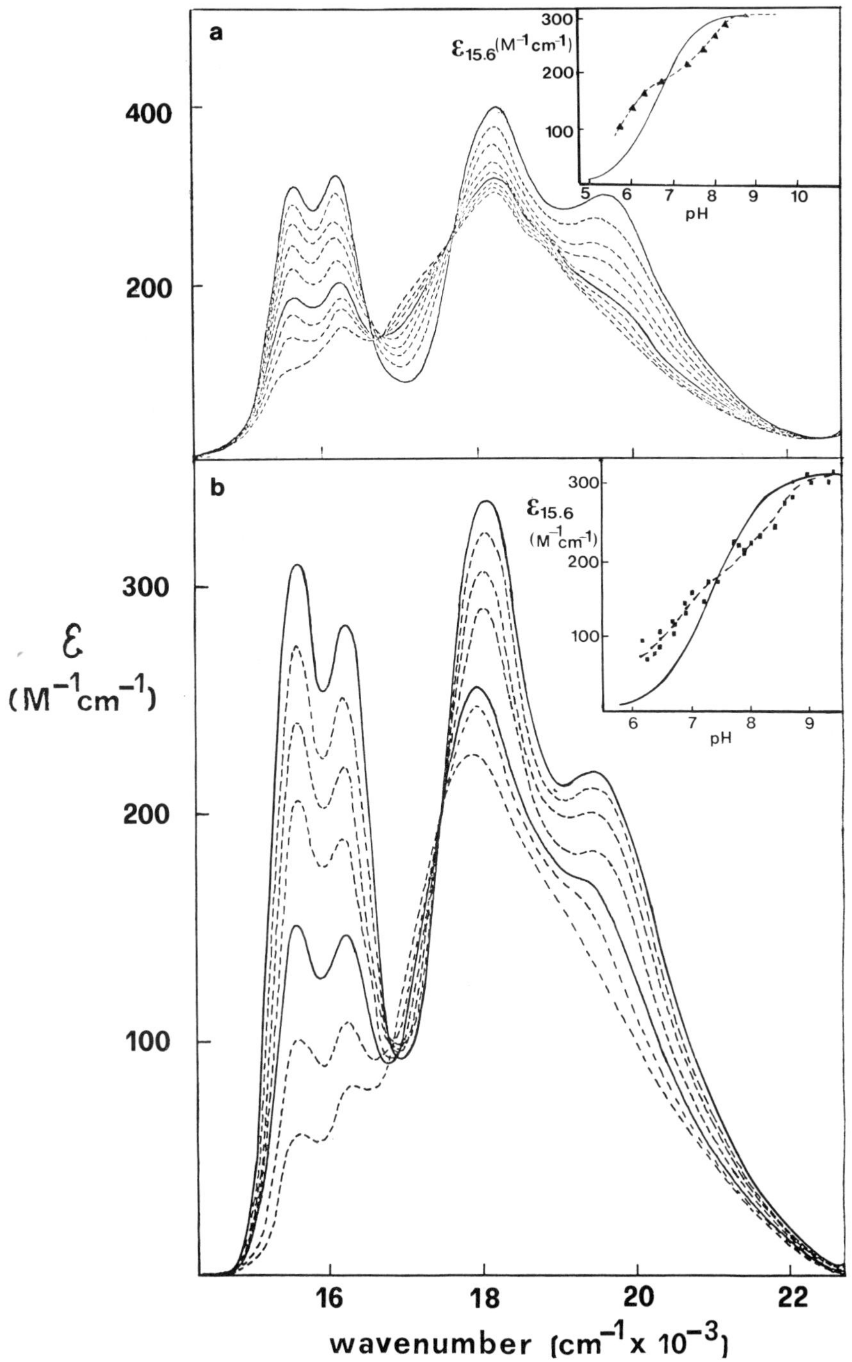

Fig. 9. Electronic absorption spectra of CoBCAB (a) and CoHCAB (b) as a function of pH. a: unbuffered solutions at pH 5.8, 6.0, 6.3, 6.7, 7.3, 7.7, 7.9, 8.2, 8.8, in order of increasing $\varepsilon_{15.6}$; b: 10^{-2} M HEPES buffered solutions at pH 6.1, 6.6, 7.1, 7.8, 8.3, 8.6, 9.5, in order of increasing $\varepsilon_{15.6}$. The *full lines* represent the spectra obtained at the middle and at the end of the titrations. The *insets* represent the intensity of the 15.6 $cm^{-1} \times 10^{-3}$ d–d transition as a function of pH. The *solid lines* are calculated assuming a single pK_a of 6.6 (a) and of 7.35 (b), respectively

These spectral changes have been assigned as due to the presence of anionic species which act as inhibitors and can be present as impurities even after exhaustive dialysis[210]. This interpretation which requires water always being coordinated as such in the enzyme finds further support in the constancy of the ^1H relaxation rates of the cobalt enzyme with pH and has been claimed to be consistent with all the experimental data available in the assumption of a single active enzyme species. The studies in our laboratory have established[224, 225] that the electronic spectra are reproducible and substantially independent of the presence of impurities and even of the buffering species, as long as they are very large molecules like HEPES (N-2-Hydroxyethylpiperazine-N′-2-ethanesulfonic acid), which presumably cannot enter the active cavity.

The spectra at pH 5.8 and 8.8 referred as the spectra of the acidic and basic forms respectively, are quite different either with respect to the energy of the absorption maxima and to the molar absorbance. Lindskog et al. demonstrated that the compound as well as its inhibitor derivatives are high spin with three unpaired electrons[227, 228]. In every case the electronic spectra show at least one absorption with $\varepsilon > 100\ M^{-1}\ cm^{-1}$ (Table 5); therefore on this basis the possibility that the chromophores are six coordinated can be ruled out[229]. The only stereochemistries consistent with these data are the pseudotetrahedral and the five coordinated[160]. The presence of several absorption maxima in the electronic spectra of the enzyme derivatives suggests also that either the tetracoordination or the five coordination occur in highly distorted chromophores. In the

Table 5. Spectral data for some cobalt(II) bovine carbonic anhydrase derivatives

Inhibitor	Band energy ($cm^{-1} \times 10^{-3}$) and molar absorptivity ($M^{-1}\ cm^{-1}$)
pure enzyme at pH 5.9	8.2(47), 10.5(13)sh, 15.6(100), 16.2(135), 18.1(250), 19.2(180)
pure enzyme at pH 8.0	8.2(80), 11.0(18), 15.6(260), 16.2(280), 18.1(380), 19.2(280)
acetate	8.3(7), 10.2(6), 14.1(9), 18.0(110), 19.4(80)sh, 21.2(100)
acetazolamide	7.9(65), 9.2(65), 11.0(45), 16.8(500)sh, 17.4(530), 18.2(220), 19.3(390)
aniline	8.9(100), 10.5(55), 17.4(470)sh, 17.9(580), 18.9(460)
azide	8.6(24), 10.3(31), 10.6(30), 15.6(65)sh, 17.6(250), 18.4(210), 21.2(110)
benzoate	8.5(15), 10.6(6), 13.2(4), 16.9(160)sh, 18.0(220), 19.7(170)sh, 20.8(110)sh
bromide	8.5(3), 10.5(7), 14.0(10)sh, 16.8(200), 18.2(200), 18.7(90)sh, 20.3(170)
chloride	8.0(14), 10.6(14), 13.9(4)sh, 16.9(270), 18.1(220), 20.2(160)
cyanate	8.0(60), 10.7(75), 17.4(600), 19.1(450), 20.4(180)sh
cyanide	7.8(100)sh, 8.7(120), 10.2(110)sh, 17.2(700), 18.4(500)sh
fluoride	7.8(15), 8.4(15), 9.0(22), 9.7(19)sh, 10.7(16), 16.9(170)sh, 18.2(270), 19.2(210)sh
formate	10.2(18), 11.0(11)sh, 13.6(15), 17.8(130), 18.2(120)sh, 18.7(110)sh, 22.0(55)sh
hydrogen carbonate	7.8(19), 8.5(21), 8.8(17), 11.0(9), 15.6(85), 16.3(105), 18.2(220), 19.4(170)sh
hydrogen sulfite	10.2(4), 13.6(8)sh, 18.2(110), 20.0(90)sh
hydrosulfide	7.8(25), 10.2(30)sh, 10.8(32), 16.3(600), 18.4(400)
iodide	10.5(1), 13.2(9), 15.9(80)sh, 16.7(120)sh, 18.1(170), 18.9(180), 20.1(190)
nitrate	8.6(16), 10.2(9), 14.2(8), 18.0(130), 21.1(100)
oxalate	14.5(14), 17.7(165), 19.3(110)sh, 21.6(100)
2,4-pentanedione	9.2(9), 11.0(7), 13.7(12)sh, 18.2(130), 20.7(90)sh
thiocyanate	10.2(4), 10.6(4)sh, 14.5(9), 17.5(100), 18.9(90)sh, 21.5(100)

Reproduced from Ref. 136

absence of rigorous information which may be obtained by relating the absorption spectra and the symmetry of the chromophore, only a thorough comparison of the electronic spectra within the series of the inhibitor derivatives may allow more reliable deductions.

The electronic spectra of the inhibitor derivatives (Table 5) are essentially pH independent although the apparent affinity constants do depend on pH (see Sect. B.III.). The spectra have been classified into three types[136]:

1. weak intensity spectra (ε max $< 150\ M^{-1}\ cm^{-1}$) with a characteristic absorption in the range 12,000–14,000 cm^{-1};
2. intense spectra with molar absorbance above 300 $M^{-1}\ cm^{-1}$ in the visible region;
3. spectra with molar absorbance in the visible region between 150 and 300 $M^{-1}\ cm^{-1}$, some of which show a weak absorption in the range 12,000–14,000 cm^{-1}.

The comparison between the first two types of spectra can lead to some conclusions. A difference in molar absorbance between 600 $M^{-1}\ cm^{-1}$ for the NCO^- derivative and 100 $M^{-1}\ cm^{-1}$ for the NCS^- derivative or between 700 $M^{-1}\ cm^{-1}$ and 110 $M^{-1}\ cm^{-1}$ for the cyanide and acetate derivatives, respectively, should be ascribed to major changes in the coordination geometry rather that to changes in the molecular orbital coefficients within the same coordination. In general, high spin tetrahedral complexes of cobalt(II) show much larger absorbance than five coordinate complexes do[229]. Furthermore the position of the absorption may be meaningful. The absorptions in the range 16000–20000 cm^{-1} which are the most intense are due to transitions from the ground level to levels arising from the ^4P term (F–P transitions) whereas the weak absorptions at lower energies are due to transitions among the levels arising from the ^4F term (F–F transitions)[136, 230–234]. Whereas the F–P transitions strongly depend on nephelauxetic effects[231–234] i.e. on electron delocalization effects, it is well established that increasing the coordination number leads to an increase in the range of the F–F transitions and therefore to an increase in energy of the highest of them[235]. Although such statements should take into account concomitant changes in the overall ligand field strength and geometrical effects, it may be reasonable to propose the occurrence of five coordination for the compounds giving rise to spectra of the type 1 and tetracoordination for the compounds giving rise to spectra of the type 2[136].

Tetracoordination should occur through substitution of a ligand of the pure cobalt enzyme and five coordination through addition of a further ligand. Reasonably in the tetracoordinated derivatives the water molecule could be removed.

The spectra of type 3 should be indicative of equilibria between tetra and five coordinated species[136, 236]:

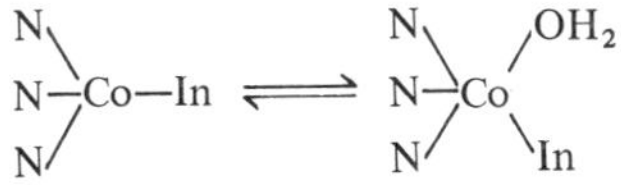

This criterion is somewhat rough since a single electronic spectrum can probably be reproduced by several combinations of ligand field and geometrical parameters; however, as long as the skeleton of the three histidine residues is assumed to be fixed the number of independent parameters is decreased. In this sense the spectral differences between the adducts with N_3^-, NCS^- and NCO^- (Fig. 10) are hardly accounted for by major angular distortions without the concomitant change in coordination number, since

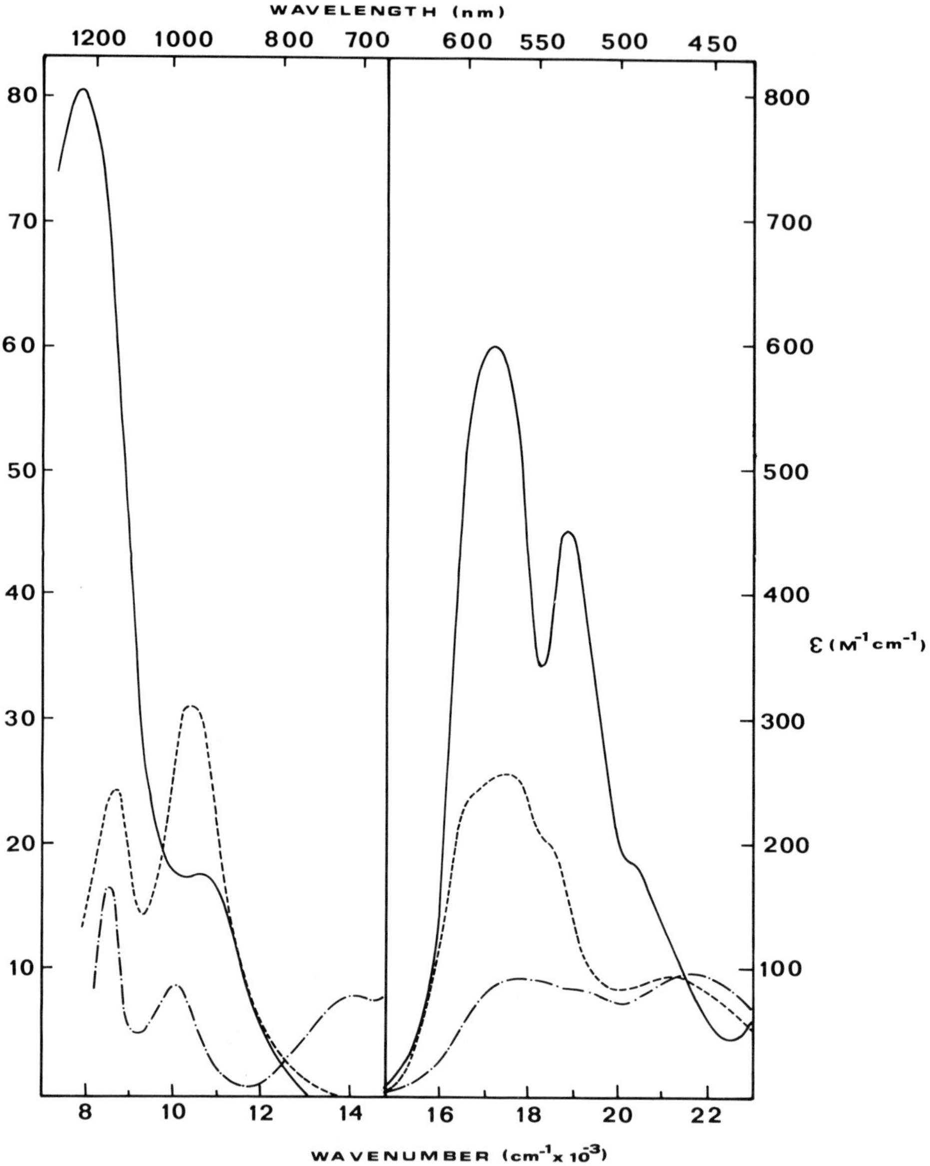

Fig. 10. Electronic absorption spectra of CoBCAB adducts with cyanate (——), azide (- - - - -), and thiocyanate (–·–·–) at pH 5.9

the three inhibitors have such similar donor and steric requirements. A further confirmation of the utility of the criterion comes from the dicyanoaurate CoBCA adduct. Using the X-ray difference map to locate the heavy gold atom, the dicyanoaurate ligand has been located in the cavity of the native enzyme in a position different from that of water[237)]. Successive studies by ^{13}C NMR on the cobalt derivatives have proved that the cyanocomplex is indeed bound to the cobalt ion[238)]. The electronic spectrum is typically of class 1 and therefore it is reasonably assigned as due to a five coordinated chromophore[238].

If the above criterion is applied to the spectra of the pure CoCA the answer is that they are indicative of a tetracoordinated geometry; however, such large spectral variations both in shape and frequency between the acidic and basic forms of the enzyme might imply some conformational change in the protein part more than a simple change of a donor atom. Indeed, the simple change of a OH_2 group with a OH^- group would be expected to only slightly affect the electronic spectra since both groups are close in the nephelauxetic and spectrochemical series[160, 240]. A change in the ligand conformation introduces a further parameter in the interpretation of the electronic spectra and makes less reliable the above criterion. The magnetic circular dicroic spectra of the cobalt enzyme have also been recorded: while the spectra of the acidic form have been interpreted as due to a pseudotetrahedral chromophore, both four and five coordination have been proposed for the alkaline form[239, 240].

^1H relaxation rate studies of water solutions containing the cobalt enzyme and its inhibitors are not of great help since CoCA shows the smallest ^1H T_1^{-1} enhancement (see

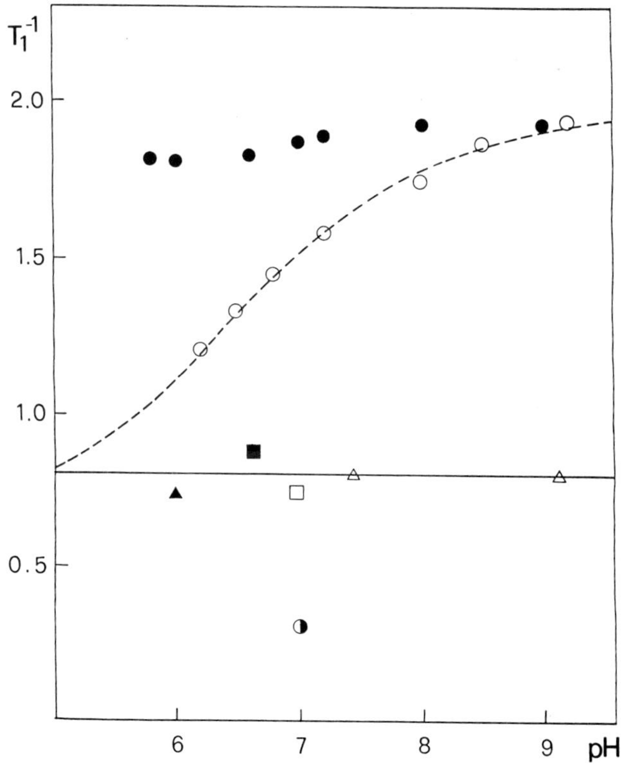

Fig. 11. ^1H T_1^{-1} values for water protons of 2.4×10^{-3} M CoBCAB unbuffered solutions as a function of pH: (●) pure enzyme; (○) plus 10^{-1} M sulfate. Residual values obtained after addition of p-toluenesulfonamide (——), cyanate (■), and thiocyanate (▲). The T_1^{-1} values of 2.4×10^{-3} M solutions of ZnBCAB (◑), $Co(H_2O)_6^{2+}$ (□), and of a five coordinated monoaqua cobalt(II) complex (△, see next section) are also shown

Fig. 6). Six coordinated high spin cobalt(II) complexes are expected, however, to give an even smaller T_1^{-1} enhancement owing to shorter electronic relaxation times[217]. Indeed, the ^1H T_1^{-1} value of a 10^{-3} M solution of cobalt sulfate containing the $Co(OH_2)_6^{2+}$ complex is as small as the values of CoCA adducts in which the bound water has been replaced by the inhibitor[136] (Fig. 11). Since all the inhibitors of CoCA reduce the ^1H T_1^{-1} values of solutions of the cobalt enzyme, even if in the case of five coordination they do not remove the bound water, it has been proposed that either the electronic relaxation times of the five coordinate species are again quite unfavourable[136, 200], or that the protons exchange slowly on the NMR time scale. Indeed, the ^1H T_1^{-1} value of a five coordinate model compound containing a water molecule ($[CoTPyMA–OH_2]^{2+}$, see next section) is the same as the residual relaxation rate (Fig. 11) (see Note Added in Proofs 2).

The EPR spectra of the cobalt enzyme derivatives have been matter of extensive investigations[138, 241–247]. However, the spectra of the pure cobalt enzyme are of very low quality probably because of the concomitant acid/base equilibria which never provide a single species. Furthermore the EPR spectra of high spin cobalt(II) compounds are difficult to interpret owing to the complexity of the theory[248]. The experiments provide three apparent g values whereas their analysis requires three molecular g values and two parameters describing the energy separation between the two Kramers doublets at zero magnetic field. Under these circumstances every attempt at theoretical interpretation is bound to fail. Furthermore even from an empirical point of view the ranges of apparent g

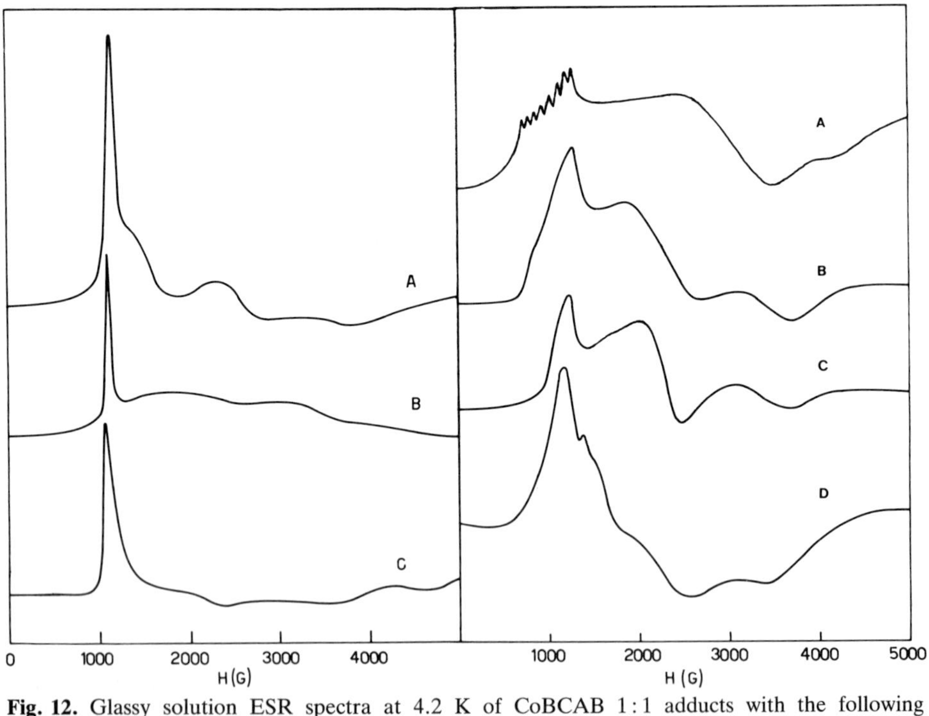

Fig. 12. Glassy solution ESR spectra at 4.2 K of CoBCAB 1:1 adducts with the following inhibitors: (*left*), A, acetazolamide, pH 8.7; B, cyanate, pH 6.1; C, cyanide, pH 8.0; (*right*), A, iodide, pH 6.0; B, thiocyanate, pH 6.2; C, acetate, pH 7.2; D, azide, pH 6.5

values for tetrahedral and five coordinate species are largely overlapping[248–255]. However, recently the EPR spectra of the 1 : 1 adducts have been recognized to belong to two types (Fig. 12) which substantially differ in the width of the low field signal (largest apparent g value)[248]. In the case of the iodide derivative a hyperfine splitting of 160×10^{-4} cm^{-1} is also resolved[248]. If the linewidth is attributed to large unresolved hyperfine coupling, the two kinds of EPR spectra can be distinguished on the basis of the different magnitude of the hyperfine coupling. The single crystal studies on cobalt complexes reported up to date allow us to propose that square pyramidal and octahedral geometries give rise to larger hyperfine splittings than tetrahedral or trigonal bipyramidal complexes[249–255]. Now, the derivatives giving rise to electronic absorption spectra of type 2, already assigned as tetrahedral, give rise to EPR spectra consistent with tetracoordinated or trigonal bipyramidal geometries whereas the derivatives giving rise to electronic absorption spectra already assigned as five-coordinated give rise to EPR spectra consistent with octahedral or square pyramidal geometries. Keeping in mind that the above geometries are only limit idealized geometries, and that deviations may be large, the EPR spectra might add further support to the above assignment of the stereochemistries of the CoCA adducts.

In the case of the cyanide adduct and in excess of inhibitor a sharp EPR spectrum is obtained which has been assigned to low spin cobalt(II) (one unpaired electron)[244–247]. Presumably two cyanide groups bind the metal ion and the resulting five coordinate chromophore is low spin. This is not unexpected since five coordinate complexes with soft donor atoms like the carbon of the cyanide ion are low spin[231]. The stability of this derivative increases with decreasing temperature: at 25 °C only a small amount is detected whereas it is completely formed in the glassy solutions below 0 °C[244]. A thorough EPR investigation with carbon and nitrogen labelled cyanide ions has allowed detection of the ligand hyperfine splitting due to two magnetically equivalent carbon atoms[246, 247]. Since a ligand hyperfine splitting due to a single nitrogen is also observed[245], a square pyramidal geometry was proposed with the unpaired electron residing in the d_z^2 orbital[244]. Such geometry could give rise to a sizeable ligand hyperfine coupling only with an axial nitrogen of the protein and undetectable couplings with the two basal nitrogens. The particularly strong C–Co bond would allow detection of sizeable couplings when such bonds occur in the basal plane of the square pyramid.

The binding of a second cyanide ligand to the cobalt enzyme in the glassy solution at low temperatures may account for the observation of an EPR spectrum assigned to high spin five coordinated species for those inhibitor derivatives giving rise to electronic absorption spectra intermediate between those assigned as tetracoordinated and those assigned as five coordinated (electronic spectra of type 3)[248]. Furthermore the EPR spectra of the cobalt enzyme in the presence of large excess of inhibitors like I^- and NCS^- or non-inhibitory substances like imidazole for the bovine enzyme are typical of high spin species different from those previously analyzed[248]. Evidently some sort of further chemical interaction with the ligands in excess within the active cavity occurs at low temperatures.

A Spectroscopic Model of Carbonic Anhydrase

In order to have a better comprehension of the pH dependent properties of carbonic anhydrase a cobalt compound with pseudotrigonal symmetry and a single water molecule has been prepared[256, 257]. The formula is $CoLH_2O(ClO_4)_2$ where L is

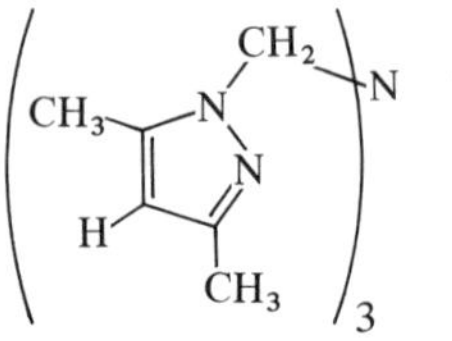

The electronic spectra in water solution are strictly similar to those in the solid state and are strongly pH dependent. Apparently the spectrum decreases in intensity with a pK_a of 7.5 until at low pH only the spectrum of the hexaaqua complex is obtained. This acid base equilibrium is believed to involve the apical nitrogen. Another equilibrium is observed with a pK_a around 9 which has been assigned to the following dissociation

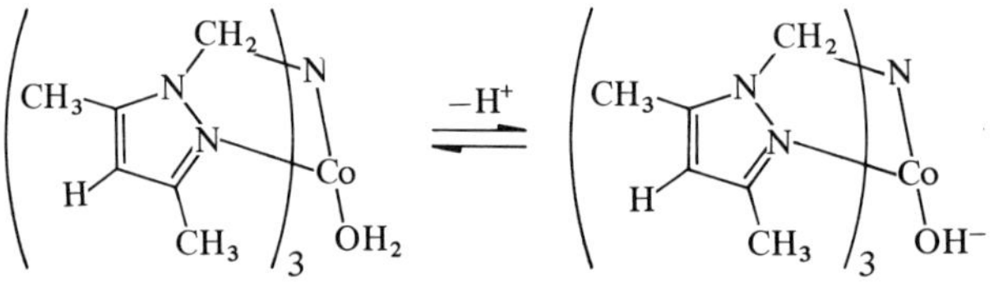

The spectral changes relative to this latter equilibrium are similar to those observed in CoCA (Fig. 13) if allowance is made for the more intense absorptions of the pseudotetrahedral cobalt enzyme and for a weak band at 13,000 cm^{-1} typical of five coordination shown by the model complex[256, 257]. The finding that the $Co{-}OH_2 \rightleftharpoons Co{-}OH$ equilibrium can dramatically affect the electronic spectrum despite the vicinity of the H_2O and OH^- donors in the spectrochemical series is a support of the bound water molecule dissociation in the enzyme.

The ligand N_3^- is capable of substituting the water in the complex, still giving rise to a five coordinate chromophore. However, its affinity constant is pH dependent, similar to the behavior observed with CA. Evidently the OH^- ion, even without cavity effects, is capable of successfully competing with the azide ion.

II. Stereochemistry of the Copper(II) Derivatives

The electronic absorption spectra of CuBCAB show a single broad absorption at 13,000 cm^{-1}. Their pH dependence is quite minor, probably with an ipsochromic shift of 300 cm^{-1} when the pH is increased[140, 180, 205, 258]. Also the EPR spectra are essentially pH insensitive with $g_\parallel = 2.31$, $g_\perp = 2.06$, $A_\parallel = 131 \times 10^{-4}$ cm^{-1} at pH 5.6, the rhombic component being small[205, 245]. Also the human isoenzyme B which shows an electronic absorption spectrum with a maximum at 13,700 cm^{-1} displays a similar behaviour although the experimental data were obtained in presence of buffering species[245, 247, 258].

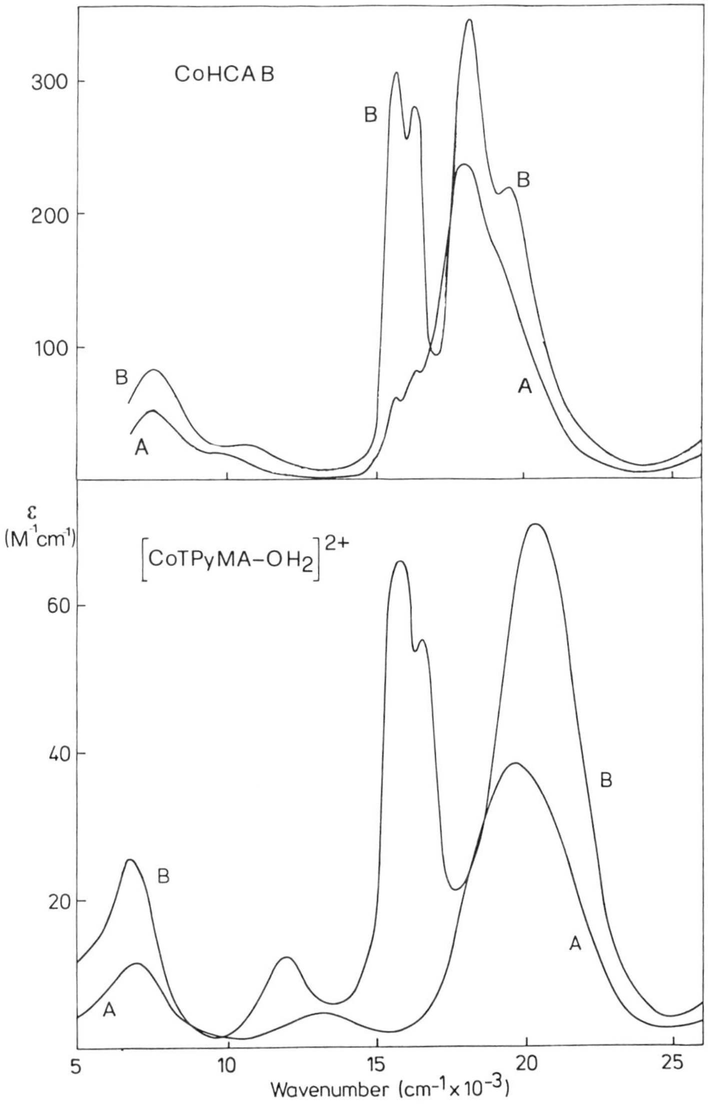

Fig. 13. Electronic absorption spectra of CoHCAB (*upper*) at pH 6.1 (A) and 9.5 (B), and of Co[tris(3.5-dimethyl-l-pyrazolylmethyl)amine $OH_2]^{2+}$ model complex (*lower*) at pH 7.5 (A) and 9.6 (B)

Despite the pH insensitivity of these spectral parameters, the apparent affinity constant of inhibitors is pH dependent, as in the native and cobalt enzyme, with a pK_a of 8.5[140, 205]. This means either that the $Cu–OH_2 \leftrightharpoons CuOH$ equilibrium hardly affects the electronic parameters (as often happens in inorganic compounds) or that the K_{app} is governed by a group inside the cavity different from H_2O and unable to affect the electronic properties.

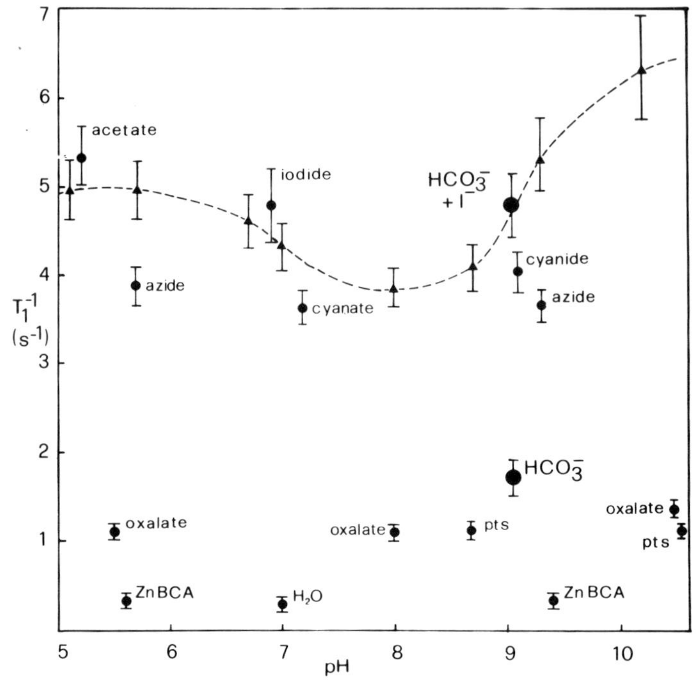

Fig. 14. ^{1}H T_1^{-1} values for water protons of 1.0×10^{-3} M CuBCAB unbuffered solutions as a function of pH: (▲) pure enzyme; (●) adducts with the indicated ligands. The T_1^{-1} values of pure water and of a 1.0×10^{-3} M solution of ZnBCAB are also shown

The limit spectra of the inhibitor derivatives are pH independent and again show a single broad absorption located between 12,000 and 14,000 cm^{-1} [140, 205]. These electronic spectra are typical of distorted chromophores either five coordinated or tetracoordinated with a flattened tetrahedral geometry[259]. Again, the choice between the two stereochemistries is based upon subtle considerations after a close inspection of a large amount of data.

In particular, evidence that inhibitors do bind the copper ion come from charge transfer absorption bands displayed by the iodide and azide derivatives[140, 205]. Furthermore ^{13}C NMR data on HCO_3^- [260] and acetate[261] have proved the existence of a copper oxygen chemical bond. Therefore it can be generalized that inhibitors do bind at the metal site. On the other hand most of the inhibitors (N_3^-, I^-, Ac^-, Fig. 14) do not remove the water molecule from coordination since they bind to the copper ion without affecting the ^{1}H NMR relaxation rate of the water solution[205]. In this case either a bound histidine residue is removed and replaced by the inhibitor or the inhibitor binds at a further coordination position. Although in principle both hypotheses are possible we would like to pragmatically assume that the histidine nitrogens are bound under all circumstances (see Note Added in Proofs 3). Within this frame the coordination of the inhibitors which do not remove water is proposed to be

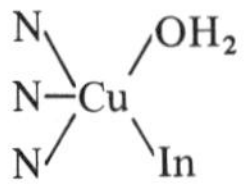

Furthermore, since the overall shapes of the electronic absorption spectra and of the EPR spectra are strictly similar along the series of compounds, one would also propose that the three nitrogen atoms and the oxygen atom of water form a flattened tetrahedral cage whose geometry remains substantially constant among the various adducts. The inhibitor could bind in a more or less loose way in the axial position as often happens for the fith donor in five coordinated copper complexes[262)]. The $A_{\parallel}$ values, which were found to be extremely sensitive to small angular variations in distorted five coordinate complexes[259, 263)], change only slightly among the adducts with the bovine isoenzyme B[205)] whereas they change somewhat more among the adducts of the human isoenzyme B[245)] (Table 6). In the series of the inhibitors of the latter isoenzyme $A_{\parallel}$ varies from 150 to 100 G with large rhombic components consistent with the above geometrical proposals. Analogously to the cobalt enzyme also in the case of the copper substituted enzyme two adducts with the cyanide ion have been characterized[247, 264)]. At room temperature, independent of the cyanide to enzyme ratio, only the 1:1 cyanide adduct is formed. At 112 °K in excess of cyanide a new species starts to appear which is completely developed at a cyanide to enzyme ratio of 6:1. This species, which has been identified as the dicyanide adduct, shows an electronic spectrum with an absorption at 16,800 cm^{-1} and an EPR spectrum with $g_{\parallel} = 2.20$, $g_{\perp} = 2.05$ and $A_{\parallel} = 185 \times 10^{-4}$ cm^{-1}. The EPR spectrum of the dicyanide adduct shows also evidence of ligand superhyperfine structure which can be attributed to two magnetically equivalent nitrogens of the protein and to two magnetically equivalent carbon ligands of the two bound cyanide groups[247)]. Whereas the 1:1 adduct, analogously to the other derivatives with anionic inhibitors, is presumably five coordinate the electronic parameters of the dicyanide derivative strongly suggest an essentially planar tetragonal chromophore, with one or two loosely bound axial

Table 6. EPR parameters for some human and bovine copper carbonic anhydrase derivatives[a]

	$g_{\parallel}$	$g_{\perp}$	$A_{\parallel}(cm^{-1} \times 10^{-4})$
CuHCAB	2.29	2.06	150
acetate	2.32	2.09	140
cyanate	2.31	2.11	119
azide	2.27	2.09	112
hydrosulfide	2.25	2.07	117
cyanide	2.24	2.09 – 2.02	122
dicyanide	2.20	2.05	190
CuBCAB	2.31	2.06	131
acetate	2.31	2.03	131
benzoate	2.26	2.02	165
oxalate	2.29	2.04	150
azide	2.26	2.04	124
cyanate	2.29	2.05	128
aniline	2.30	2.02	131
imidazole	2.27	2.04	164
nitrite	2.30	2.03	134

[a] At liquid nitrogen temperature. The EPR parameters of CuHCAB derivatives are taken from Ref. 245 with the exception of the cyanide adduct which is taken from Ref. 247. Those of CuBCAB are taken from Ref. 205

ligands[205, 247, 259]. The planar moiety could be formed by two histidines and two cyanide ions.

The oxalate ion being potentially bidentate has also been tested against CuBCAB[205]. ^1H T_1 measurements have shown that oxalate reduces the T_1^{-1} enhancement, whereas the EPR and electronic spectra are similar to those of the other anionic derivatives. Although it may be possible that the ligand prevents the bound water from fast exchange, the bidentate nature of the ligand leads to the proposal that water is removed from coordination and that the structure is

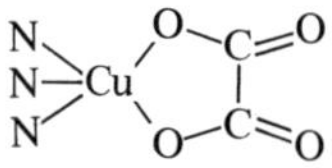

HCO_3^- and some sulfonamides probably remove water from coordination (at least for the bovine enzyme B) since they cause a decrease of the ^1H relaxation rate (Fig. 14)[205, 260]. It should be recalled here that sulfonamides are the strongest inhibitors of the native enzyme and that HCO_3^- is the natural substrate. The chromophores they give rise to are therefore

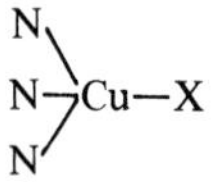

X = HCO_3^-, sulfonamides

Sulfanilamide and *p*-toluenesulfonamide have a different behavior with the human and bovine isoenzymes[265]. In the former case the ^1H relaxation rate is not reduced, and the EPR spectra of the adducts show $A_{\parallel}$ values in the range 130–140 $\times 10^{-4}$ cm^{-1}. The same ligands with the bovine B enzyme quench the ^1H relaxation enhancement and show $A_{\parallel} = 70 \times 10^{-4}$ cm^{-1}. Such low $A_{\parallel}$ values have supported the assignment of the chromophores as a pseudotetrahedral geometry[205, 265].

The free copper enzyme and the HCO_3^- derivative have electronic parameters similar to those of the other anionic inhibitor derivatives. This may not be a worry if it is considered that the inhibitor is the fifth ligand loosely bound, but it may also indicate that in the free copper enzyme there is a fifth donor which could be either a second water molecule or a donor provided by the protein part. In the latter case a reasonable candidate could be the threonine 199 residue which is hydrogen bonded to the coordinated water in the native protein[46]. The two accessible binding sites present in the cavity both in the cobalt and copper derivative probably partially overlap. Indeed, the iodide causes detachment of HCO_3^- from the copper derivative[260], although it binds to a different site. Only in the case of the strong and small inhibitor cyanide two ions are capable of binding the metal at low temperature.

III. Stereochemistry of the Oxovanadium(IV) Derivative

The oxovanadium(IV) ion differs from the other ions in that the metal ion is always accompanied by the oxygen as an extra ligand. Through EPR investigations the occur-

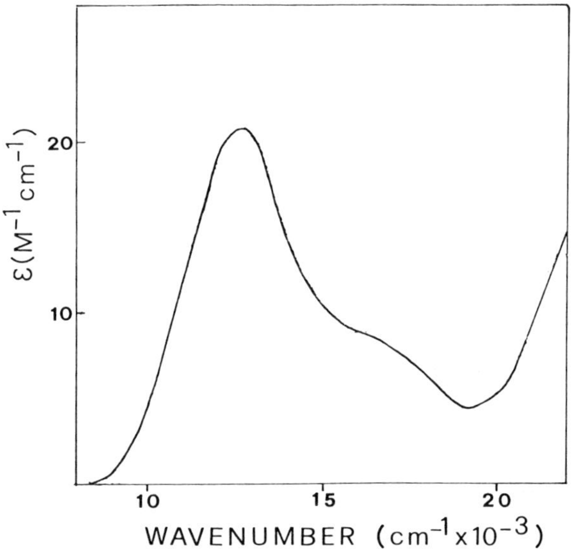

Fig. 15. Electronic absorption spectrum of oxovanadium(IV) bovine carbonic anhydrase B at pH 6.0

rence of two forms of the oxovanadium(IV) substituted enzyme have been shown[191]. The low pH form is the dominant one at pH values below 7, the pK_a of this equilibrium being estimated to be 7.7. Both the electronic spectra (Fig. 15)[206] and the EPR spectra[191] have been interpreted on the basis of a tetragonal symmetry either square pyramidal or elongated octahedral[206]. The compound apparently oxydizes above pH 8[191]. Since from ^1H NMR data a water molecule seems still bound to the metal the proposed coordination of vanadium(IV) is

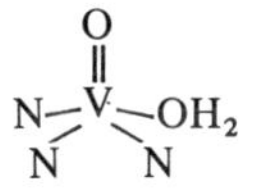

the VN_3 moiety being puckered owing to the steric requirements of the protein nitrogens. The possibility for the threonine hydroxide to occupy the sixth coordination position cannot be ruled out. Indeed, from the ESR spectra an increase of the coordination number in the acidic form as compared to that of the alkaline form has been proposed, the further donor being a protein group[191]. Inhibitors like NCS^- and N_3^- seem not to bind to the metal ion[193].

IV. Stereochemistry of the Nickel(II) Derivatives

The electronic spectra of NiBCA show absorptions at 15.6 and 25.6 × 10^3 cm^{-1}, with molar absorbances of 20 and 50 M^{-1} cm^{-1}, respectively[143]. The energies are typical of six coordination; although the intensities are known to be sensitive to the lack of symmetry of the chromophore they are higher than the usual values[266, 267]. The spectra are also pH dependent: by increasing the pH the intensity of the former absorption band first

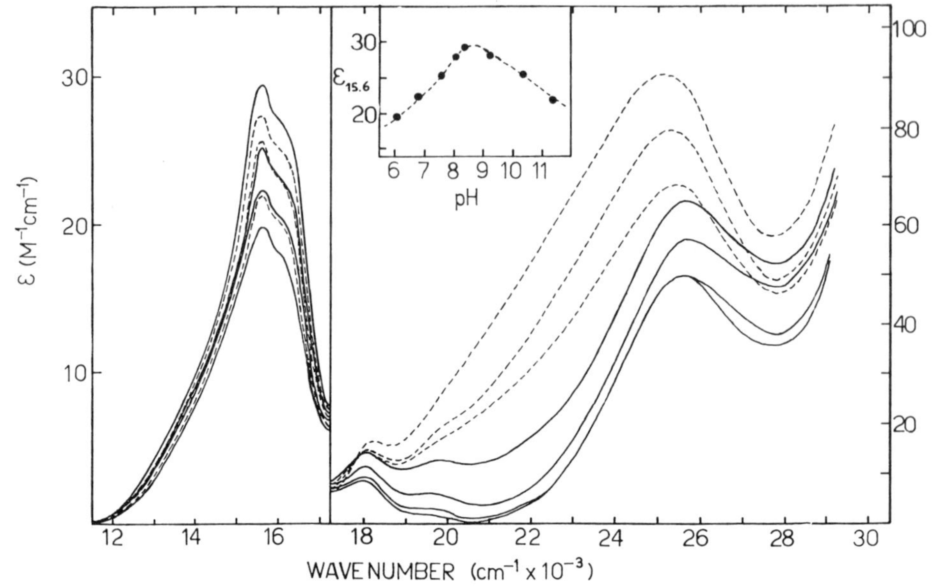

Fig. 16. Electronic spectra of NiBCAB; spectra at pH = 6.1, 6.8, 7.6, and 8.4 are represented by *solid lines* in order of increasing absorbance, whereas spctra at pH = 9.3, 10.4, and 11.4 are represented by *dashed lines* in order of decreasing absorbance at 15.6×10^3 cm^{-1} and of increasing absorbance at 25×10^3 cm^{-1}. The *inset* shows the pH dependence of $\varepsilon_{15.6}$

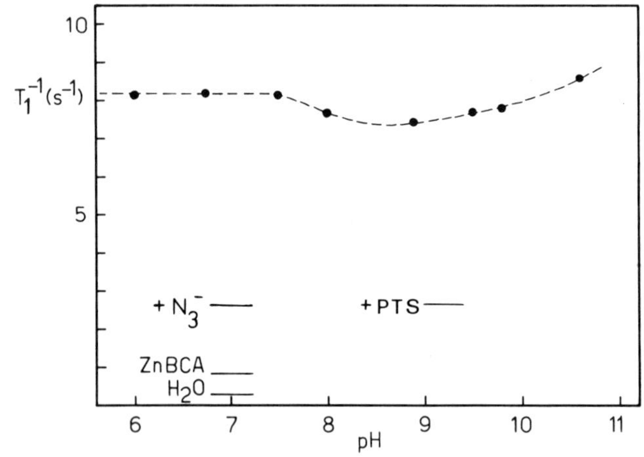

Fig. 17. ^1H T_1^{-1} values for water protons of 5.0×10^{-3} M NiBCAB unbuffered solutions as a function of pH (●). Residual values obtained after addition of azide or *p*-toluenesulfonamide, as well as the T_1^{-1} values of pure water and of 5.0×10^{-3} M ZnBCAB are also shown

increases and then decreases (Fig. 16) indicating that more than one acid-base equilibrium occurs. The six coordination may be reached through the three histidine nitrogens, two water molecules and a sixth donor group which can be either a third water molecule or the threonine OH group[193]. The binding of a N_3^- ion occurs in a 1 : 1 ratio, and large water $^1H\ T_1^{-1}$ enhancement remains upon addition of N_3^-[143] (Fig. 17). Therefore more than one water molecule is indeed bound to the paramagnetic center.

V. *Stereochemistry of the Manganese(II) Derivatives*

The studies of MnCA are rather numerous owing to the large relaxing capabilities of the manganese(II) ion on nearby nuclei[141, 142, 190, 196, 201, 204, 268, 269]. However, the stability of MnCA is rather small and decreases at low pH values, so that at pH 7 a 10^{-3} M MnCA solution contains about 20% of free manganese ions[190, 201].

The manganese ion in MnCA is high spin[141] and therefore the electronic absorption spectra are not meaningful from the point of view of the stereochemistry. The manganese CA is among the few manganese proteins which allow the observation of EPR spectra[141, 190, 268]; however, the correlation between the EPR parameters and the metal stereochemistry is a difficult task[268]. Again the largest source of information is the water ^1H relaxation rate enhancement which shows the presence of a water molecule within the coordination sphere[196, 201, 204]. The inhibitor N_3^- does not remove the water from coordination, thus giving rise to five coordination (Fig. 18). The oxalate ion and *p*-toluenesul-

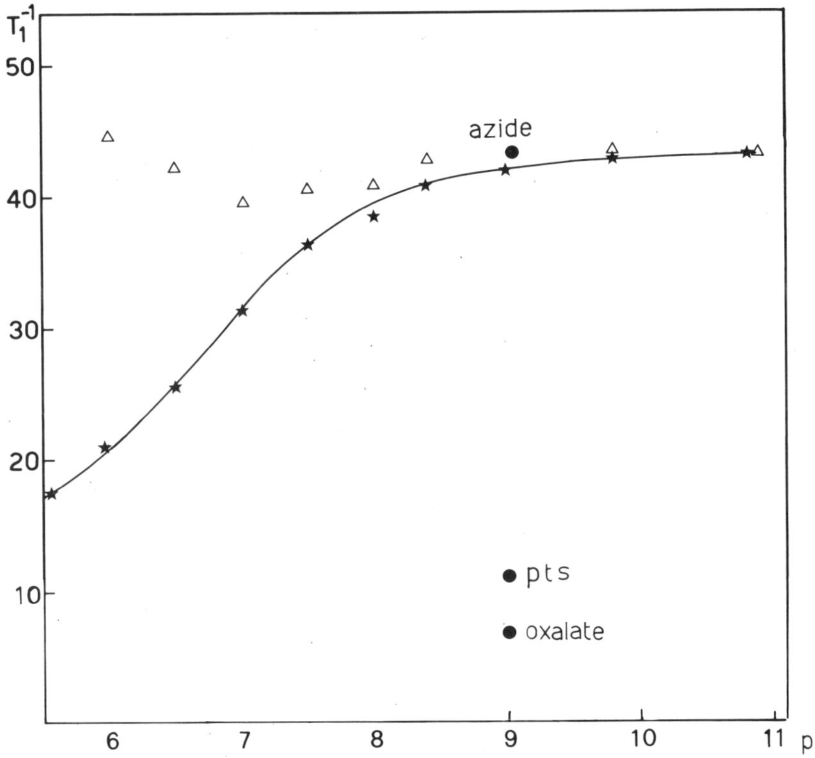

Fig. 18. $^1H\ T_1^{-1}$ values for water protons of 1.0×10^{-3} M MnBCAB unbuffered solutions as a function of pH: (★) experimental data; (△) data corrected for manganese dissociation at low pH; (●) adducts with the indicated ligands

fonamide, analogously to CuCA, do remove the water molecule[204)]. The non inhibited enzyme may be either four or five coordinated.

A ^{1}H NMR investigation of N^{1}-acetylsulfanilamide interacting with MnCA permits mapping of the molecule within the active site. The estimated metal-proton distances fit with a model in which the sulfonamide nitrogen is directly bound to the metal[141)].

VI. Stereochemistry of the Cobalt(III) Derivatives

The cobalt(III) carbonic anhydrase is obtained by oxidizing the cobalt(II) enzyme with hydrogen peroxide[192, 270)]. The resulting product is diamagnetic indicating that the cobalt(III) ion inside the active site is six coordinated[270)]. Also the electronic absorption spectrum, which shows a maximum at 19,400 cm^{-1}, is in agreement with a distorted octahedral coordination of the cobalt(III) ion. The donor set is probably similar to that of the nickel derivative. Addition of anions like CN^-, ClO_4^-, and N_3^- causes the above band to shift, indicating a direct binding at the metal.

VII. Stereochemistry of the Dichloromercury(II) Derivative

Mercury dichloride forms a compound with apo CA. On the basis of the X-ray data it seems that the linear Cl–Hg–Cl moiety bends until an angle of *circa* 90° is formed and the coordination of the mercury atom is completed by three histidyl nitrogens and the hydroxide of the threonine 199 residue[193)]. The reciprocal orientation of the donor atoms as they appear from the X-ray analysis is shown in Fig. 19.

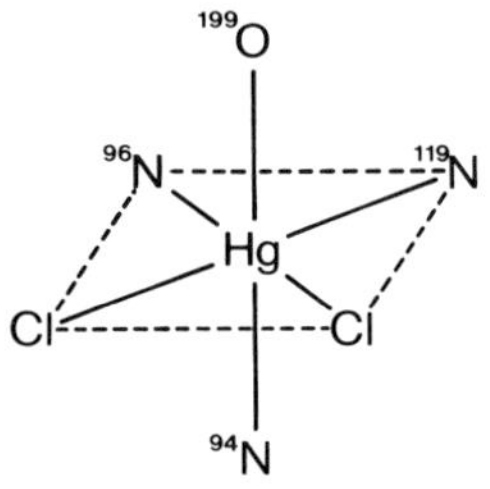

Fig. 19. Coordination polyhedron of mercury(II) in carbonic anhydrase. The protein ligands are His 94, His 96, His 119, and Thr 199

VIII. Stereochemistry of the Cadmium(II) Derivative

The substitution of the native zinc with a cadmium(II) ion leads to an enzyme which still shows esterase activity[195)]. The ^{111}CdHCAB has been estensively studied through perturbed angular correlation of γ rays (PAC)[139, 195, 271)]. These studies showed the existence of two pH dependent forms of the cadmium enzyme, the interconversion from one to the other occurring with a $pK_a = 9.0$, in close agreement with esterase activity measurements[195)]. The observed change in the quadrupole parameters which can be obtained through the PAC technique have been shown to be consistent with the ionization of a cadmium bound water molecule. Coordination stereochemistries analogous to those of the native zinc enzyme have been proposed for CdCA as well as for the imidazole and sulfonamide derivatives[139].

^{113}CdHCAB and ^{113}CdBCAB have been also investigated through ^{113}Cd NMR spectroscopy[272–274]. In the case of the ^{113}CdHCAB derivative no signal was formerly[272, 273]. observed at pH $< pK_a$ whereas at pH above 9.1 or in the presence of inhibitors like Cl^-, Br^-, and I^-, a resonance became detectable. These observations were explained on the basis of relatively slow exchange processes occurring at the cadmium atom in the enzyme. Owing to the large range of chemical shifts shown by ^{113}Cd in different environments (dramatic differences in chemical shifts have been in fact observed in the ^{113}Cd derivatives of the various isoenzymes[272]), the linewidths of the ^{113}Cd resonances are expected to be modulated by relatively slow exchange processes which generally do not affect the linewidth of the NMR signals of the other commonly investigated nuclei[272, 273]. These slow processes were proposed to involve conformational rearrangements, or the equilibrium itself between the acidic and the alkaline forms of the enzyme which the authors considered as due to water dissociation[272, 273]. Recently[274], also the low pH form of the human B isoenzyme has been reported to give rise to a detectable resonance, provided that the solution is free from carbon dioxide and buffering species.

F. The Affinity of the Inhibitors of the Native Enzyme for the Metal Substituted Carbonic Anhydrases

As mentioned in the previous sections, the inhibitors of the native CA isoenzymes are generally capable of interaction with the metal substituted derivatives. The affinity of the inhibitors for the native enzyme is most often obtained from activity measurements whereas physical chemical measurements are used for measuring the affinity of the various molecules for the various metalloenzymes.

The affinity values of inhibitors for the cobalt enzyme are often of the same order of magnitude as for the native enzyme (Table 2). Sometimes some compounds are found through spectroscopic techniques to interact with the metal ion of CoCA and they are tacitally included among the inhibitors of the enzymatic activity of both the native and cobalt(II) substituted derivatives. In such a way a large number of compounds have been tested in order to point out the requirements of the molecule to act as inhibitor. Furthermore, the spectroscopic techniques applied to the systems CoCA-inhibitors have allowed an accurate study of the pH dependence of the apparent affinity constants of the various inhibitors on an extended range of pH values even in absence of buffers. The conclusions of such results presumably hold also for the native enzyme.

The affinity constants of the anions for CuCA are much larger than for the native enzyme whereas they are practically zero for the oxovanadium derivative. The affinity of aromatic sulfonamides drops for non active metallo derivatives. Table 2 is a survey of the literature data obtained through different techniques including the data for the zinc derivative.

Anionic inhibitors have been shown through activity measurements to have pH dependent affinity constants for the native enzyme, the affinity decreasing with increasing pH[47, 48, 68, 69, 144]. The same behavior is shown by the cobalt(II)[131, 137, 145] and the other

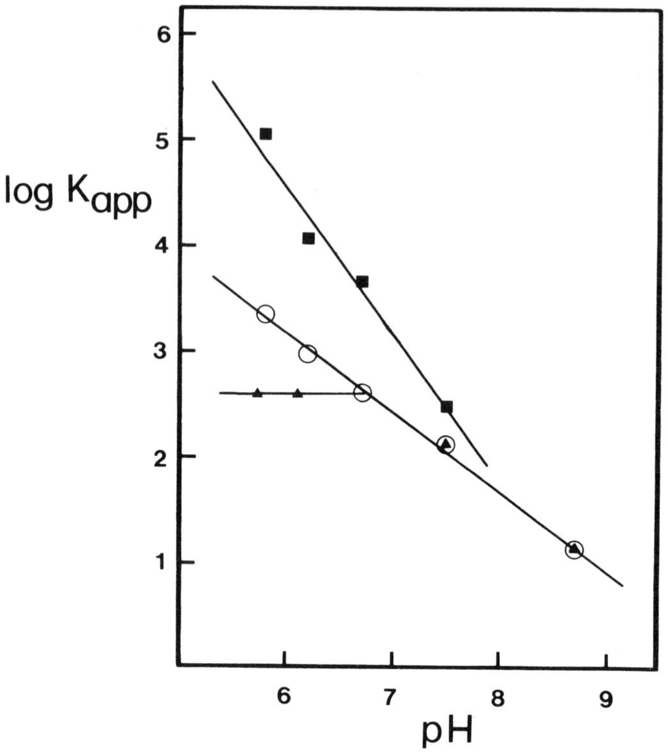

Fig. 20. pH dependence of the apparent affinity constants of acetate and oxalate ions for CoBCA: (▲) acetate in 0.1 M phosphate buffer; (○) acetate in unbuffered solutions; (■) oxalate in unbuffered solutions

metal derivatives[139, 140, 205], indicating the presence in all cases of at least one acid-base equilibrium in the active site responsible for the binding of inhibitors. Actually, for some of the metal derivatives the pH dependence of the affinity constant is the only proof of the existence of such equilibria since the spectral variations with pH of the pure metalloenzymes are far less dramatic than in the case of cobalt. However, the above ionizing group may in principle be different from that governing the electronic spectra of CoCA and from that governing the catalytic activity of the enzyme. As a matter of fact, more than one acid-base equilibrium are capable of influencing the binding of inhibitors, as observed in the case of carboxylate and bicarboxylate anions interacting with cobalt carbonic anhydrase[145]: the binding is more and more favoured as the pH is decreased, far beyond what it would be expected in the case of a single ionization being operative (Fig. 20). ^{1}H and ^{13}C NMR measurements have also shown the existence of a second binding site for the acetate ion in the active cavity, different from the metal ion but close enough to it to influence the ligand nuclear relaxation parameters; it seems that this second site can be occupied also in the presence of a sulfonamide molecule bound to the metal ion[142, 144, 275, 276].

Zwitterionic substances like aminoacids show[277] a bell shaped pH dependence of the apparent affinity towards CoBCAB (Fig. 21) similar to that of CN^-, HS^- or sul-

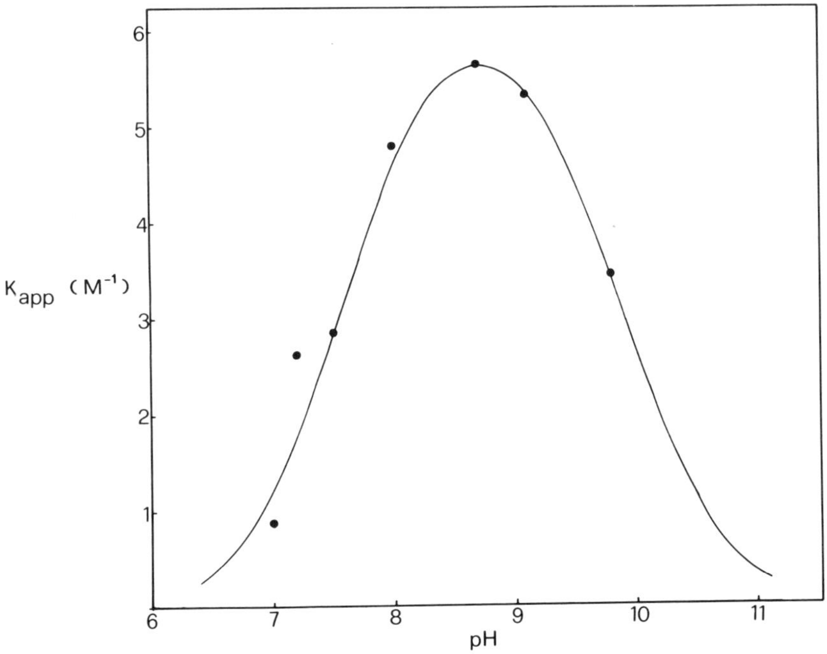

Fig. 21. pH dependence of the apparent affinity constant of L(+)alanine for CoBCA. The *full line* is fitted to Eq. 2, Sect. B.III.)

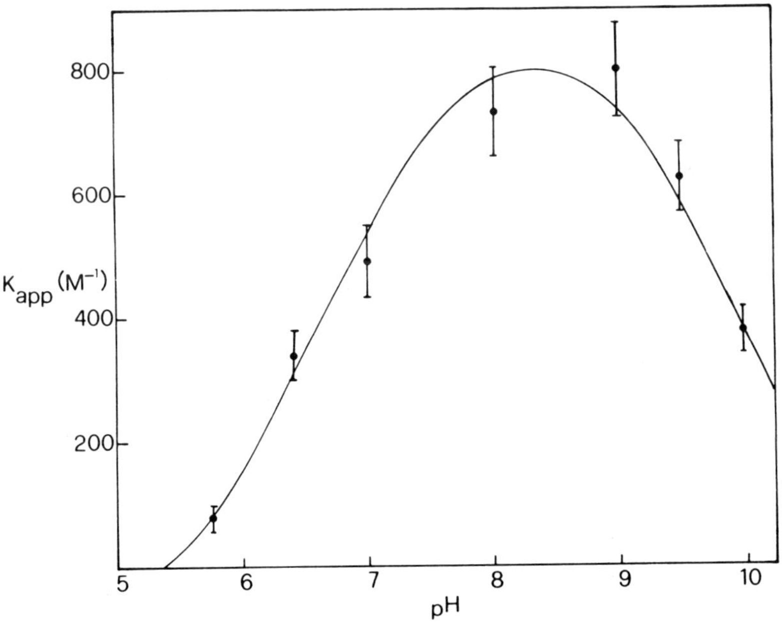

Fig. 22. pH dependence of the apparent affinity constant of trichloracetaldehyde for CoBCAB. (The *full line* is fitted to Eq. 2, Sect. B.III.)

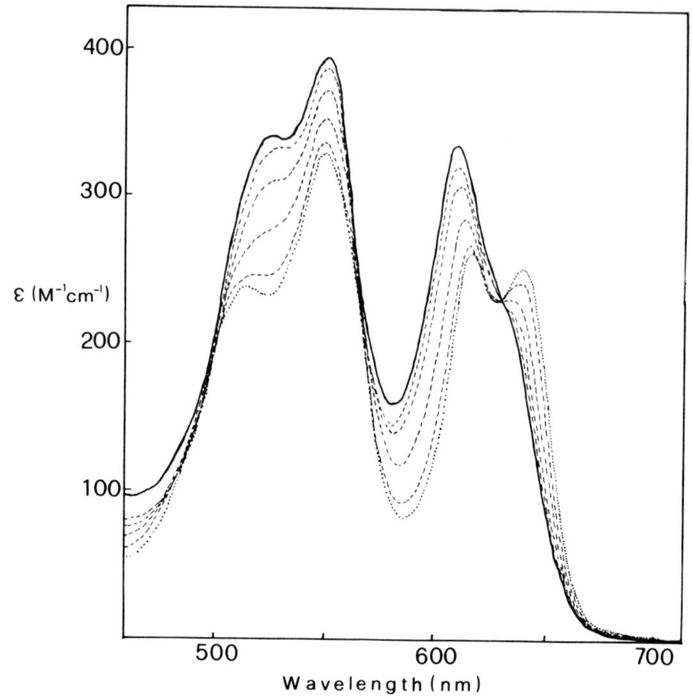

Fig. 23. Electronic absorption spectra of CoBCAB at pH 8.0, alone (.....) and at various trichloroacetaldehyde concentrations (-----). The *solid line* approximates the limit spectrum of the adduct within 3%

fonamides (see Sect. B. III.). Also in this case it has been proposed[277] that the interacting species are the anionic form of the inhibitors with the acidic form of the enzyme in analogy with the behaviour of the anions.

Neutral inhibitors are usually weak ligands, with some exceptions which are worthy of notice. Aniline shows the same pH dependence as anions[143], both from the point of view of its inhibiting properties towards the zinc enzyme[171] and of its binding affinity towards the cobalt derivative[169, 170]; therefore the negative charge of the ligand seems not to be essential for binding to the positively charged acidic form of the enzyme (see Note Added in Proofs 4), although it contributes to enhance the affinity. Hydrated trichloroacetaldehyde, which is also of interest representing the product of the enzyme catalyzed hydration of trichloroacetaldehyde, behaves with respect to pH like sulphonamides and aminoacids (Fig. 22)[278]. Once again the bell shaped dependence of the affinity constant on pH can be rationalized in terms of the ionization of the aldehyde with a $pK_a = 10.0$. The electronic spectra of the adduct with the cobalt enzyme have the peculiarity of being similar to those of the high pH form of the enzyme[278] (Fig. 23), at variance with those of all the other inhibitor derivatives which never show the characteristic bands between 600 and 640 nm. A further proof of direct binding at the metal came from ^{1}H and ^{13}C NMR measurements, which definitely show the ligand to be within binding distance to the metal ion[278].

Methanol has also been reported to be a weak inhibitor of the zinc enzyme with inhibiting capability increasing with increasing pH[68, 171]; however, successive measurements on the cobalt derivative showed that the metal is not involved in binding, since the electronic spectra are unaffected and the NMR relaxation parameters of the methanol nuclei indicate a distance of about 600 pm from the metal ion[170)]. It is possible that methanol occupies the same binding site as that found in the case of the acetate ion for the second binding site[276)].

All the inhibitors mentioned up to now are non competitive with respect to the substrates of the enzymatic reactions catalyzed by both human and bovine isoenzymes. Actually, no competitive inhibitor of carbonic anhydrase was known since 1971, when an extensive report appeared on the inhibiting properties of imidazole towards HCAB[102)]. This substance, however, had no measurable effect on the bovine enzyme. It was soon apparent that imidazole had some unique properties: U. V. difference spectra indicated a different binding mode with respect to the other inhibitors[279)], while its affinity for the enzyme seemed to be almost pH independent[139, 279)], unless the pH was lowered below the pK_a of the ligand protonation; X-ray data also suggested that the imidazole ring occupied a different position in the active site, being at about 270 pm from the metal and bent enough to leave the water molecule coordinated to the metal[179)]. Consistently,

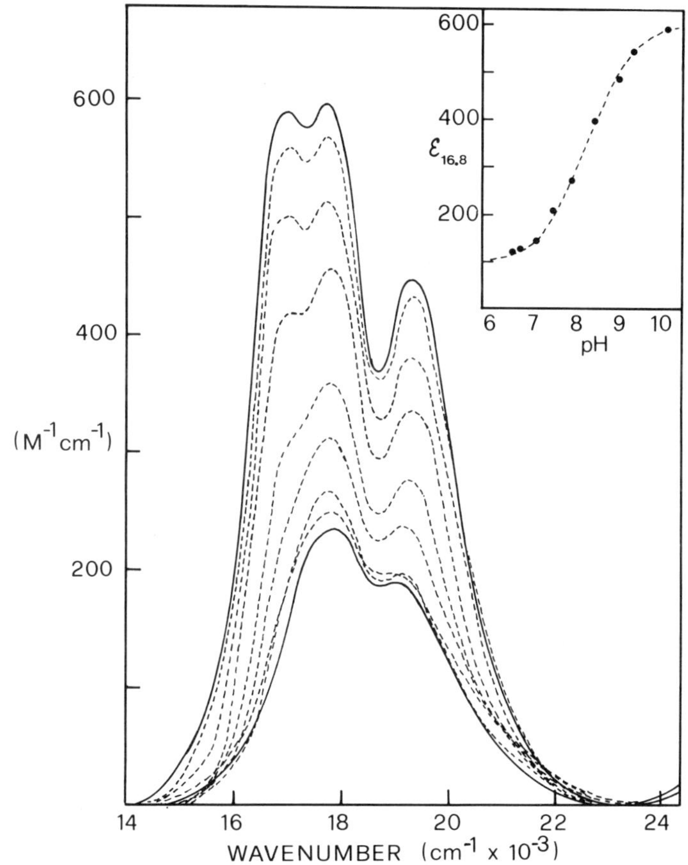

Fig. 24. pH dependence of the electronic spectra of the imidazole adduct of CoHCAB. From the bottom pH 6.5, 6.65, 7.05, 7.4, 7.9, 8.4, 9.0, 9.3, 10.1

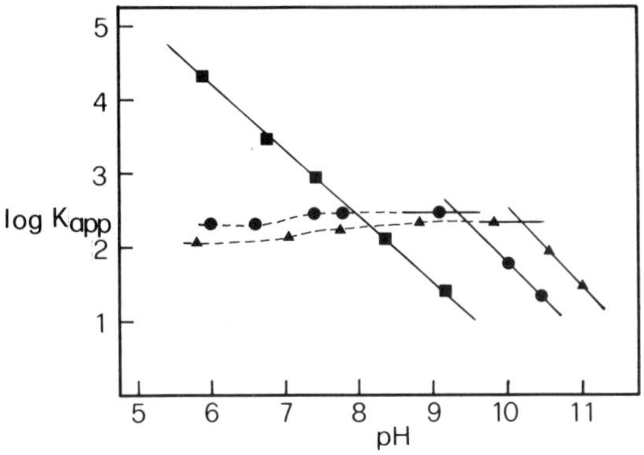

Fig. 25. pH dependence of the apparent affinity constants of 1, 2, 4 triazole (▲), 1, 2, 3 triazole (●), and tetrazole (■) for CoBCAB. The pK_a of the pyrrolic protons are 10.3, 9.4, and 4.9, respectively

Perturbed Angular Correlation spectra obtained on the cadmium derivative[139)] lead to the proposal of five coordination around the metal ion, although the electronic spectra of the cobalt derivative were more indicative of pseudotetrahedral geometry; finally, the electronic spectra of the imidazole adduct of CoHCAB are strongly pH sensitive (Fig. 24), while its stability constants are not[139, 280)]. A recent attempt to clarify the binding behaviour of the above ligand has lead to the discovery of analogues with much higher affinity towards the enzyme, namely 1, 2, 3 triazole, 1, 2, 4 triazole and tetrazole[280)]. The above substances are also capable of binding to the bovine enzyme CoBCAB, although with lower affinity constants than in the case of the human enzyme; the affinity constants of all of them are pH independent, up to the ligand NH ionization (Fig. 25). This new class of CA inhibitors should be characterized by the possibility of a bridging interaction between the metal ion and some residue in the active site capable of forming hydrogen bonds with the NH group of the ligand. On the basis of the X-ray structure of the imidazole adduct the residue is tentatively proposed to be Gln 92. A strong interaction with another site in the cavity would account for the insensitiveness of the affinity constant to the metal-linked ionizations.

G. Substrates and Metal Substituted Carbonic Anhydrases

Metal substitution has been also of primary importance in shedding some light on the type of interactions between substrates and the active site cavity. The largest amount of data is of course referred to carbon dioxide and to the hydrogenocarbonate ion, which are the natural substrates of the enzyme. From kinetic measurements on the native enzymes (see Sect. B. II.) it has been established that a specific binding site is required for both CO_2 and HCO_3^-. ^{13}C NMR spectra of $H^{13}CO_3^-$ at pH 8.3 in the presence of cobalt bovine carbonic anhydrase established through an analysis of the nuclear relaxation times that the anion is actually bound to the paramagnetic metal[281, 282)]; discrepancies in the calculated affinity constants with respect to those obtained through anion competition were attributed to the presence of a second, non catalytic binding site[281)].

It would be obviously desirable to obtain the same kind of information on $^{13}CO_2$; unfortunately, the pK_a of the equilibrium $CO_2 + H_2O \leftrightharpoons HCO_3^- + H^+$ is 6.35, and therefore the pH cannot be lowered so much to completely avoid the presence of HCO_3^- in solution, owing to possible metal dissociation or denaturation of the protein. The ^{13}C NMR spectra of the system $CO_2 \rightleftharpoons HCO_3^-$ in water show two signals separated by ~40 ppm[282, 283], but addition of even small amounts of the high activity bovine cobalt enzyme causes collapse of the signals to an average position[281] with consequent loss of specific information on the single species. In the case of the low activity cobalt human enzyme B the interconversion rate between CO_2 and HCO_3^- is slower, and two separate signals are still present at enzyme concentrations suitable for the experiment; however, the interconversion is still too fast with respect to the intrinsic relaxation rates of ^{13}C of CO_2 and HCO_3^- in the presence of the paramagnetic center. Therefore T_1 and T_2 give once again only an average information[282].

Recently the same kind of experiment was performed on the copper derivative[260] which displays much lower activity as well as longer electronic relaxation times, both favourable properties in order to obtain different relaxation times for the two species. As a matter of fact, a dramatic difference was observed in the paramagnetic effect on $H^{13}CO_3^-$ and $^{13}CO_2$, which establishes that the former ion is bound to the metal ion while the latter species is not, but has a different binding site in the cavity about 600 pm far from the copper center.

Furthermore, the dissociation rate of HCO_3^- from the copper coordination sphere could be measured, the resulting value being at least two orders of magnitude smaller than the corresponding value for the zinc enzyme[260]. This relatively slow rate is consistent with the above dissociation being the slow step in the catalytic pathway for the copper enzyme, accounting for the reduced activity of the copper derivative.

At variance with all the monoanionic ligands of the copper derivative, HCO_3^- reduces the water ^1H T_1^{-1} enhancement[260], although it apparently competes with anions for the same or nearly the same coordination site (Fig. 14).

Trichloroacetaldehyde has been reviewed in the preceeding section, being a non competitive inhibitor of the esterase activity. However, it can be also considered as a product of the enzyme catalyzed aldehyde hydration, since it is nearly 100% in the hydrated form in water solution. An investigation by means of spectroscopic techniques[278] has shown that the above substance is bound to the metal ion in CoBCAB. Analogously, NMR experiments indicated the presence of paramagnetic coupling between the nuclei of acetaldehyde and the cobalt ion[126]. The above experiments establish that the CoBCAB catalyzed hydration of acetaldehyde follows the same pH dependence observed for the zinc enzyme.

If it is recalled that the carboxylate anions are products of ester hydrolysis, and that they have long been known to bind metallosubstituted carbonic anhydrases, it may be suggested that the products of every catalyzed reaction can bind the metal at some step of the catalytic process. Things are less clear as far as the substrates of the hydration reaction are concerned: only carbon dioxide has been shown not to bind the metal ion[260] while no direct evidence exist for esters and anhydrous aldehydes, although they probably behave like the natural substrate.

It can now be attempted to summarize the progress obtained on the knowledge of the active site chemistry referring to a schematic representation of a possible catalytic pathway (Fig. 26). For the sake of simplicity and without loss of generality the alkaline form of the enzyme is drawn as a tetrahedral zinc hydroxo complex (A). A zinc aquo complex could also fit in the scheme, provided that a proton transfer to another basic group in the

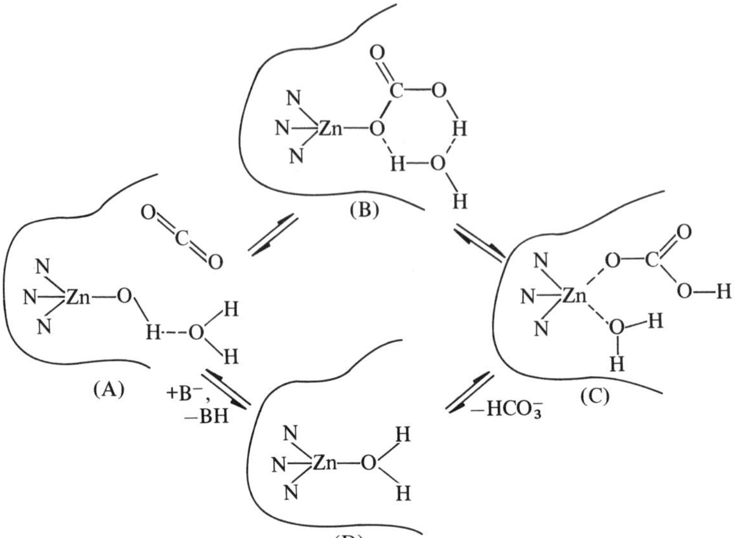

Fig. 26. Schematic representation of a possible catalytic pathway, summarizing the structural information obtained from physico-chemical data

cavity is made possible. Consistently with the NMR data, the incoming CO_2 molecule approaches the active site possibly in the hydrophobic part of the cavity, without binding to the metal. It could, however, bind through H-bond any other attacking site in the vicinity of the metal. A possible candidate for this may be the Gln 92 residue to which the competitive inhibitor imidazole is proposed to bind[260]. By reacting with the hydroxide, the bicarbonate ion *is* coordinated to the metal ion (B). Such a reaction may occur through several pathways, e.g. by involving a water molecule of the cavity. At this stage the metal can coordinate a water molecule as a fifth ligand (C), giving rise to a five coordinate intermediate to facilitate the detachment of the HCO_3^- ion (D). The presence of an equilibrium between a four and a five coordinate HCO_3^- adduct is consistent with the spectral data of the cobalt derivative[136]. The alkaline form of the enzyme is finally reconstituted by a possible buffer mediated proton transfer.

H. Conclusions

The substitution of the native zinc(II) ion with other metal ions has allowed to gain information on some key properties of the enzyme. In particular, the dichloromercury(II) derivative has shown the flexibility of the active site with respect to its coordinating properties, in that also an OH threonine residue may act as a sixth coordination ligand. Such a group is a good candidate as a donor atom also for other six coordinated complexes like nickel(II), cobalt(III), and possibly oxovanadium(IV). In such cases the metal ion would be somewhat displaced from the native zinc position, and this could be checked through difference electron density maps.

Regarding the groups determining the pH dependent properties of the enzyme, the electronic spectra of the cobalt(II) derivative have been shown to be dependent on more than one acid-base equilibrium. Previous potentiometric studies[284–286] have shown that in the range 5–8 there are about four titratable protons. ^{113}Cd showed that at least one acidic group is capable of affecting the cadmium NMR spectra, suggesting that the ionizing group is directly bound to the metal ion.

^{1}H NMR studies of the various paramagnetic metal derivatives have shown that a group with exchangeable protons is always bound to the metal. While at low pH there is little doubt that the above group is a water molecule, the existence of an OH group at alkaline pH may be consistent with the data by assuming a shorter metal oxygen distance. The similarity of the spectral properties of the couple $CoN_4(H_2O)^{2+}$-$CoN_4(OH)^+$ in a model complex with those of the cobalt enzyme makes the above assumption less speculative[256, 257].

The whole set of physico-chemical measurements on the paramagnetic metal ion derivatives has shown that anionic ligands can either substitute the water molecule or add to a fifth coordination site. In the cavity therefore there are two binding sites, partially overlapping. This finding supports the possibility of five coordinated intermediates occurring along the catalytic pathway. Monte Carlo calculations have been performed by Clementi and coworkers[287–289] using the atomic coordinates of the active site residues in HCAC as obtained by X-ray data, a water molecule as a probe, and the interaction potentials of water with the zinc(II) ion and the various aminoacid residues[290, 291]. These studies have shown that besides the potential well in which the coordinated water molecule is present there are two more potential wells, less defined and located at about 220 and 290 pm from the zinc ion. These potential wells are not necessarily occupied but their existence, besides supporting the hypothesis of the two binding sites, may also account at least in part for the observation of residual paramagnetic effects on proton relaxation when the water molecule is removed from coordination.

NMR techniques and electronic spectroscopy established that all the products of the catalytic reactions (HCO_3^-, hydrated aldehydes and carboxylate anions) do bind the metal ion, while the substrates probably do not; their binding site(s) are not well identified so far.

Many ligands, especially if negatively charged, bind the metal in MCA. The acid-base equilibria occurring in the cavity determine the apparent affinity constants at every pH. Binegatively charged ligands may have lower affinity than mononegative analogues, as in the case of the $H_2PO_4^-$-HPO_4^{2-} system[292], or higher affinity when chelating behavior is allowed[145]. Hydrophobic (sulfonamides) and hydrophylic (imidazole) interactions may also be operative; in their absence, neutral or zwitterionic species[277] are unfavoured with respect to anionic ligands with similar steric requirements. Such requirements play an important role by themselves, as shown[238] by the large affinity displayed by the linear $Au(CN)_2^-$ ion compared with the non appreciable affinity of the bulkier planar $Au(CN)_4^-$ species.

I. References

1. Maren, T. H.: Physiol. Rev. *47*, 595 (1967)
2. Tashian, R. E. et al.: In: The Isozymes, vol. 4, p. 207; (ed. Markert, C. L.) New York: Academic Press 1975
3. Lindskog, S. et al.: In: The Enzymes, vol 5, p. 587; (ed. Boyer, P. D.) New York: Academic Press 1971
4. Coleman, J. E.: In: Inorganic Biochemistry, vol 1, p. 488; (ed. Eichorn, G. L.) Amsterdam: Elsevier 1973
5. Wyeth, P., Prince, R. H.: Inorganic Perspectives in Biology and Medicine *1*, 37 (1977)
6. Pocker, Y., Sarkanen, S.: Advan. Enzymol. *47*, 149 (1978)
7. Carter, M. J.: Biol. Rev. *47*, 465 (1972)
8. Polya, J. B., Wirtz, A. J.: Enzymologia *28*, 355 (1965); ibid. *29*, 27 (1965)
9. Addink, A. D. F.: In: Some Aspects of Carbonic Anhydrase of Sepia Officinalis(L), Drukkerij, Elinkwijk, Utrecht, 1968
10. Sanders, F., Maren, T. H.: Mol. Pharmacol. *3*, 204 (1967)
11. Brundell, J., Falkbring, S. D., Nyman, P. O.: Biochim. Biophys. Acta *284*, 311 (1972)
12. Waygood, E. R.: In: Methods in Enzymology, Vol. 2, p. 836 (1955)
13. Rossi, C., Chersi, A. Cortivo, M.: In: NASA (Natl. Aeron. Space Admin.) Spec. Publ., NASA SP-188, 1969, 131
14. Pocker, Y., Ng, S. Y. J.: Biochemistry *12*, 5127 (1973); ibid: *13*, 5186 (1974)
15. Atkins, C. A., Patterson, B. D., Graham, D.: Plant. Physiol. *50*, 214 (1972)
16. Pocker, Y., Miksch, R. R.: Biochemistry *17*, 1119 (1978)
17. Bundy, H. F.: Comp. Biochem. Physiol. *57 B*, p. 1 (1977)
18. Keilin, D., Mann, T.: Biochem. J. *34*, 1163 (1940)
19. Scott, D. A., Mendive, J. R.: J. Biol. Chem. *139*, 661 (1941)
20. Peterman, M. L., Hakala, N. V.: J. Biol. Chem. *145*, 701 (1942)
21. Keilin, D., Mann, R.: Nature *144*, 442 (1939)
22. Nyman, P. O.: Biochim. Biophys. Acta *52*, 1 (1961)
23. Rickli, E. E., Edsall, J. T.: J. Biol. Chem. *237*, 258 (1962)
24. Laurent, G. et al.: C. R. Soc. Biol. *156*, 1461 (1962)
25. Rickli, E. E. et al.: J. Biol. Chem. *239*, 1065 (1964)
26. Laurent, G. et al.: Bull. Soc. Chem. Biol. *47*, 1101 (1965)
27. Lindskog, S.: Biochim. Biophys. Acta *39*, 218 (1960)
28. Tashian, R. E. et al.: In: Biophysics and Physiology of Carbon Dioxide (ed. Bauer, C., Gros, G., Bartels, H.) Springer, Berlin, p. 165, 1980
29. Anderson, B., Nyman, P. O., Strid,L.: Biochem. Biophys. Res. Commun. *48*, 670 (1972)
30. Marriq, C., et al.: Biochimie *55*, 1361 (1973)
31. Lin, K. T. D., Deutsch, H. F.: J. Biol. Chem. *248*, 1885 (1973)
32. Giraud, N., Marriq, C., Laurent-Tabusse, G.: Biochimie *56*, 1031 (1974)
33. Henderson, L., Henriksson, D., Nyman, P. O.: Biochem. Biophys. Res. Commun. *52*, 1388 (1973)
34. Henderson, L. E., Henriksson, D., Nyman, P. O.: J. Biol. Chem. *251*, 5457 (1976)
35. Lin, K. T. D., Deutsch, H. F.: J. Biol. Chem. *249*, 2329 (1974)
36. Sciaky, M. et al.: C. R. Acad. Sci. Paris. *279 D*, 1217 (1974)
37. Gulian, M. M. et al.: Biochimie *59*, 293 (1977)
38. Gulian, M. M. et al.: Biochimie *60*, 473 (1978)
39. Henderson, L. E.: In: Biophysics and Physiology of Carbon Dioxide, (ed. Bauer, C., Gros, G., Bartels, H.) Springer, Berlin, p. 154, 1980
40. Kannan, K. K.: ibid., p. 184
41. Fridborg, K. et al.: J. Mol. Biol. *25*, 505 (1967)
42. Kannan, K. K. et al.: Cold Spring Harbor Symp. *36*, 221 (1971)
43. Kannan, K. K. et al.: J. Mol. Biol. *63*, 601 (1972)
44. Kannan, K. K. et al.: Proc. Natl. Acad. Sci. USA *72*, 51 (1975)
45. Waara, I. et al.: In: Adv. Expl. Met. Biol., Hemoglobin and Red Cell Structure and Function (ed. Brewer, G. J.), Plenum Press, New York, Vol. 28, p. 169, 1972

46. Liljas, A. et al.: Nature New Biol. *235*, 131 (1972)
47. Nostrand, B., Waara, I., Kannan, K. K.: In: The Isozymes (ed. Markert, C. L.), Academic Press, New York, p. 575, 1975
48. Waara, I.: Inaugural Dissertation, University of Uppsala, Sweden, 1974
49. Foveau, D., Sciaky, M., Laurent, G.: C. R. Sér. D. *278*, 959 (1973)
50. Whitney, P. L.: Eur. J. Biochem. *16*, 126 (1970)
51. Khalifah, R. G.: Biochemistry *16*, 2236 (1977)
52. Khalifah, R. G. et al.: Biochemistry *16*, 2241 (1977)
53. Jeffers, P. K., Sutherland, W. Mc I., Khalifah, P. G.: Biochemistry *17*, 1305 (1978)
54. Schoot Uiterkamp, A. J. M. et al.: Biochemistry *17*, 3730 (1978)
55. Whitney, P. L., Nyman, P. O., Malmström; B. G.: J. Biol. Chem. *242*, 4206 (1967)
56. Bradbury, S. L.: J. Biol. Chem. *244*, 2002 (1969)
57. Bradbury, S. L.: J. Biol. Chem. *244*, 2010 (1969)
58. Göthe, P. O., Nyman, P. O.: FEBS Lettrs *21*, 159 (1972)
59. Cybulsky, D. L. et al.: J. Biol. Chem. *248*, 3411 (1973)
60. Henkart, P., Darner, F.: J. Biol. Chem. *246*, 2714 (1971)
61. Meldrum, N. U., Roughton, F. J. W.: J. Physiol. (Lond.) *75*, 4 p and 15 p; Proc. Roy. Soc. (Lond.), B *111*, 296
62. Pocker, Y., Meany, J. E., Davis, B. C.: Biochemistry *13*, 1411 (1974)
63. Tashian, R. E., Plato, C. C., Shows, T. B.: Science *140*, 53 (1963)
64. Malmström, B. G. et al.: Eur. Biochem. Soc. Symp., I, "Structure and Activity of Enzymes" (ed. Goodwin, T. W., Harris, J. T., Hartley, B. S.), Academic Press, New York, p. 121
65. Pocker, Y., Stone, J. T.: J. Am. Chem. Soc. *87*, 5497 (1965)
66. Pocker, Y., Stone, J. T.: Biochemistry *6*, 668 (1967)
67. Armstrong, J. M. et al.: J. Biol. Chem. *241*, 5137 (1966)
68. Veerporte, J. A., Mehta, S., Edsall, J. T.: J. Biol. Chem. *242*, 4221 (1967)
69. Thorslund, A., Lindskog, S.: Eur. J. Biochem. *3*, 117 (1967)
70. Lindskog, S., Thorslund, A.: Eur. J. Biochem. *3*, 453 (1968)
71. Pocker, Y., Storm, D. R.: Biochemistry *7*, 1202 (1968)
72. Pocker, Y., Beug, M. W.: Biochemistry *11*, 698 (1972)
73. Pocker, Y., Watamori, N.: Biochemistry *10*, 4843 (1971)
74. Pocker, Y., Watamori, N.: Biochemistry *12*, 2475 (1973)
75. Pocker, Y. et al.: J. Am. Chem. Soc. *100*, 2883 (1978)
76. Pocker, Y., Guilbert, L. J.: Biochemistry *13*, 70 (1974)
77. Schneider, F., Lieflander, M.: Z. Physiol. Chem. *334*, 279 (1963)
78. Pocker, Y., Sarkanen, S.: Biochemistry *17*, 1110 (1978)
79. Whitney, P. L. et al.: J. Biol. Chem. *242*, 4212 (1967)
80. Henkart, P., Guidotti, G., Edsall, J. T.: J. Biol. Chem. *243*, 2447 (1968)
81. Henkart, P., Dorner, F.: J. Biol. Chem. *246*, 2714 (1971)
82. Lo, K. W., Kaiser, E. T.: Chem. Commun. *834* (1966)
83. Coleman, J. E.: J. Biol. Chem. *243*, 4574 (1968)
84. Kaiser, E. T., Lo, K. W.: J. Am. Chem. Soc. *91*, 4912 (1969)
85. Kaiser, E. T.: Acc. Chem. Res. *3*, 145 (1970)
86. Pocker, Y., Meany, J. E.: J. Am. Chem. Soc. *87*, 1809 (1965)
87. Pocker, Y., Meany, J. E. Biochemistry *4*, 2535 (1965)
88. Pocker, Y., Dickerson, D. G.: Biochemistry *7*, 1995 (1968)
89. Pocker, Y., Meany, J. E.: Biochemistry *6*, 239 (1967)
90. Pocker, Y., Meany, J. E.: J. Am. Chem. Soc. *89*, 631 (1967)
91. Edsall, J. T., Wyman, J.: In: Biophysical Chemistry, Vol. 1, p. 550, Academic Press, New York, 1958
92. Edsall, J. T.: In: NASA (Natl. Aeron. Space. Admin.), Spec. Publ. NASA SP, 188, 15 (1969)
93. Lindskog, S., Coleman, J. E.: Proc. Nat. Acad, Sci. USA *70*, 2505 (1973)
94. Sirs, J. A.: Trans. Faraday Soc. *54*, 201 (1957)
95. Gibbons, B. H., Edsall, J. T.: J. Biol. Chem. *238*, 3502 (1963)
96. Kern, D. M.: J. Chem. Educ. *37*, 14 (1960)
97. Ho, C., Sturtevant, J. M.: J. Biol. Chem. *238*, 3499 (1963)
98. Gibbons, B. M., Edsall, J. T.: J. Biol. Chem. *239*, 2539 (1964)

99. Kernohan, J. C.: Biochim. Biophys. Acta *81*, 346 (1964)
100. Kernohan, J. C.: Biochim. Biophys. Acta *96*, 304 (1965)
101. Pocker, Y., Bjorkquist, D. W.: Biochemistry *16*, 5698 (1977)
102. Khalifah, R. G.: J. Biol. Chem. *246*, 2561 (1971)
103. Koenig, S. H. et al.: Pure Appl. Chem. *40*, 103 (1974)
104. De Voe, H., Kistiakowsky, G. B.: J. Am. Chem. Soc. *83*, 274 (1961)
105. Khalifah, R. G.: In: Biophysics and Physiology of Carbon Dioxide (ed. Bauer, C., Gros, G., Bartels, H.), Springer, Berlin, p. 206, 1980
106. Lindskog, S.: ibid., p. 230
107. Simonsson, I., Jonsson, B. H., Lindskog, S.: Eur. J. Biochem. *93*, 409 (1979)
108. Magid, E.: Biochim. Biophys. Acta *151*, 236 (1968)
109. Pocker, Y.: In: Solute Solvent Interactions (ed. Bertini, I., Dei, A., Lunazzi, L.), London: Plenum Publishing Company, in the press
110. Koenig, S. H., Brown III, R. D.: Proc. Natl. Acad. Sci. USA *69*, 2422 (1972)
111. Strader, D. J., Khalifah, R. G.: J. Am. Chem. Soc. *98*, 5043 (1976)
112. Khalifah, R. G., Edsall, J. T.: Proc. Natl. Acad. Sci. USA *69*, 172 (1972)
113. Steiner, H., Jonsson, B. H., Lindskog, S.: Eur. J. Biochem. *59*, 253 (1975)
114. Steiner, H., Jonsson, B. H., Lindskog, S.: FEBS Lett. *62*, 16 (1976)
115. Jonsson, B. H., Steiner, H., Lindskog, S.: FEBS Lett. *64*, 310 (1976)
116. Whitney, P. L., Brandt, H.: J. Biol. Chem. *251*, 3862 (1976)
117. Khalifah, R. G.: Proc. Natl. Acad. Sci. USA *70*, 1986 (1973)
118. Prince, R. H., Wolley, P. R.: Bioorg. Chem. *2*, 337 (1973)
119. Pocker, Y., Tanaka, N.: Science *199*, 907 (1978)
120. Jonsson, B. H., Steiner, H., Lindskog, S.: Abstracts 10th Meeting of the Federation European Biochemical Societies, 1975, Abstr. 784
121. Silverman, D. N., Tu, C. K.: J. Am. Chem. Soc. *97*, 2263 (1975)
122. Silverman, D. N. et al.: J. Am. Chem. Soc. *101*, 6734 (1979)
123. Tu, C. K., Silverman, D. N.: J. Am. Chem. Soc. *97*, 5935 (1975)
124. Silverman, D. N., Tu, C. K.: J. Am. Chem. Soc. *98*, 978 (1976)
125. Silverman, D. N., Tu, C. K., Wynns, G. C.: In: Biophysics and Physiology of Carbon Dioxide (ed. Bauer, C., Gros, G. Bartels, H.), Springer, Berlin, p. 254, 1980
126. Cheshnovsky, D., Navon, G.: Biochemistry *19*, 1866 (1980)
127. Pocker, Y., Stone, J. T.: Biochemistry *7*, 2936 (1968)
128. Kernohan, J. C.: Biochim. Biophys. Acta *118*, 405 (1966)
129. Lanir, A., Navon, G.: Biochemistry *10*, 1024 (1971)
130. Whitney, P. L.: Anal. Biochem. *57*, 467 (1974)
131. Taylor, P. W., Feeney, J., Burgen, A. S. V.: Biochemistry *10*, 3866 (1971)
132. Coleman, J. E.: Nature *214*, 193 (1967)
133. King, R. W., Burgen, A. S. V.: Proc. R. Soc. Lond. B *193*, 107 (1976)
134. Maren, T. H., Wiley, C. E.: J. Med. Chem. *11*, 228 (1968)
135. Lindskog, S., J. Biol. Chem. *238*, 945 (1963)
136. Bertini, I. et al.: J. Am. Chem. Soc. *100*, 4873 (1978)
137. Lindskog, S.: Biochemistry *5*, 2641 (1966)
138. Cockle, S. A., Lindskog, S., Grell, E.: Biochem. J. *143*, 703 (1974)
139. Bauer, R., Limkilde, P., Johansen, J. T.: Carlsberg Res. Commun. *42*, 325 (1977)
140. Morpurgo, L. et al.: Arch. Biochem. Biophys. *170*, 360 (1975)
141. Lanir, A., Navon, G.: Biochemistry 11, 3536 (1972)
142. Lanir, A., Navon, G.: Biochim. Biophys. Acta *341*, 75 (1974)
143. Bertini, I., Borghi, E., Luchinat, C.: Bioinorg. Chem. *9*, 495 (1978)
144. Lanir, A., Navon, G.: Biochim. Biophys. Acta *341*, 65 (1974)
145. Bertini, I., Luchinat, C., Scozzafava, A.: Bioinorg. Chem. *9*, 93 (1978)
146. Mann, T., Keilin, D.: Nature *146*, 164 (1940)
147. Maren, T. H., Parcell, A. L., Malik, M. N.: J. Pharmacol. Exp. Theor. *130*, 389 (1960)
148. Maren, T. H. et al.: Biochem. Pharmacol. *6*, 21 (1960)
149. Coleman, J. E: a) Annu. Rev. Pharmacol. *15*, 221 (1975); b) J. Biol. Chem. *243*, 4574 (1968)
150. Bayne, W. F., Chu, L. C., Theeuwes, F.: J. Pharm. Sci. *7*, 912 (1979)
151. Taylor, P. W., King, K. W., Burgen, A. S. V.: Biochemistry *9*, 3894 (1970)

152. Petersen, R. L. et al.: Biochemistry *16*, 726 (1977)
153. Carey, P. R., King, R. W.: Biochemistry *13*, 2843 (1979)
154. Kakeya, N. et al.: Chem. Pharm. Bull. *17*, 1010 (1969)
155. Chen, R. F., Kernohan, J. C.: J. Biol. Chem. *242*, 5813 (1967)
156. Mushak, P., Coleman, J. E.: J. Biol. Chem. *247*, 373 (1972)
157. Kumar, K., King, R. W., Carey, P. R.: Biochemistry *15*, 2195 (1976)
158. Bindford, J. S., Lindskog, S., Wadsö, I.: Biochim. Biophys. Acta *341*, 345 (1974)
159. Williams, R. P. J.: Proc. Colloq. Protides Biol. Fluids *14*, 25 (1967)
160. Dennard, A. F., Williams, R. J. P.: In: Transition Metal Chemistry, vol. 2, p. 115 (ed. Carlin, R. L. Marcel Dekker), New York
161. Riepe, M. E., Wang, J. H.: J. Biol. Chem. *243*, 2779 (1968)
162. Feeney, J., Burgen, A. S. V., Grell, E.: Eur. J. Biochem. *34*, 107 (1973)
163. Ward, R. L.: Biochemistry *8*, 1879 (1969); ibid.: *9*, 2447 (1970)
164. Ward, R. L., Cull, M. D.: Arch. Biochem. Biophys. *150*, 436 (1972)
165. Ward, R. L.: In: Biophysics and Physiology of Carbon Dioxide (ed. Bauer, C., Gros, G., Bartels, H.), Springer, Berlin, p. 262, 1980
166. Ward, R. L., Whitney, P. L.: Biochim. Biophys. Res. Commun. *51*, 343 (1973)
167. Brown, G. S., Navon, G., Shulman, R. G.: Proc. Natl. Acad. Sci. USA *74*, 1794 (1977)
168. Appleton, D. W., Sarkar, B.: Proc. Natl. Acad. Sci. USA *71*, 1686 (1974)
169. Bertini, I., Luchinat, C., Scozzafava, A.: J. Am. Chem. Soc. *99*, 581 (1977)
170. O'Connor Westerik, J.: J. Biol. Chem. *253*, 8499 (1978)
171. Appleton, D. W., Sarkar, B.: Bioinorg. Chem. *4*, 309 (1975)
172. Wolpert, H. R., Strader, C. D., Khalifah, R. G.: Biochemistry *16*, 5717 (1977)
173. Campbell, I. D., Lindskog, S., White, A. I.: J. Mol. Biol. *98*, 597 (1975)
174. Pesando, J. M., Grollman, A. P.: Biochemistry *14*, 689 (1975)
175. Harrowfield, J., Mac, B., Norris, V., Sargeson, A. M.: J. Am. Chem. Soc. *98*, 7282 (1976)
176. Gupta, R. K., Pesando, J. M.: J Biol. Chem. *250*, 2630 (1975)
177. Pesando, J. M.: Biochemistry *14*, 681 (1975)
178. Pocker, Y., Stone, J. T.: Biochemistry *7*, 4139 (1968)
179. Kannan, K. K. et al.: FEBS Lett. *73*, 115 (1977)
180. Lindskog, S., Nyman, P. O.: Biochim. Biophys. Acta *85*, 462 (1964)
181. Hunt, J. P., Rhee, M. J., Storm, C. B.: Anal. Biochem. *55*, 615 (1977)
182. Kidani, Y., Hirose, J., Koiche, H.: J. Biochem. *79*, 43 (1976)
183. Kidani, Y., Hirose, J.: J. Biochem. *81*, 1383 (1977)
184. Henkens, R. W., Watt, G. D., Sturtevant, J. M.: Biochemistry *8*, 1874 (1969)
185. Tupper, R., Watts, R. W. E., Wormall, A.: Biochem. J. *50*, 429 (1952)
186. Henkens, R. W., Sturtevant, J. M.: J. Am. Chem. Soc. *90*, 2669 (1968)
187. Henkens, R. W.: Biochemistry *8*, 1874 (1969)
188. Lindskog, S., Malmström, B. G.: J. Biol. Chem. *237*, 1129 (1962)
189. Kidani, Y., Hirose, J.: Chem. Letters *475* (1977)
190. Wilkins, R. G., Williams, K. R.: J. Am. Chem. Soc. *96*, 2241 (1974)
191. Fitzgerald, J. J., Chasteen, N. D.: Biochemistry *13*, 4338 (1974)
192. Shinar, H., Navon, G.: Biochim. Biophys. Acta *334*, 471 (1974)
193. Kannan, K. K.: Structure and Function of Carbonic Anhydrases. In: Proceedings on Biomolecular Structure, Conformation, Function and Evolution (ed. Srinivasan, R.), Pergamon Press, England (1978)
194. Lindskog, S.: Struct. Bonding *8*, 153 (1970)
195. Bauer, R., Limkilde, P., Johansen, J. T.: Biochemistry *15*, 334 (1976)
196. Lanir, A., Gradstajn, S., Navon, G.: Biochemistry *14*, 242 (1975)
197. Mildvan, A. S., Cohn, M.: Advan. Enzymol. *33*, 1 (1970)
198. Dwek, R. A.: Advan. Mol. Relax. Proc. *4*, 1 (1972)
199. Burton, D. R.: In: ESR and NMR of Parmagnetic Species in Biological and Related Systems (ed. Bertini, I., Drago, R. S.), NATO-ASI, D. Reidel Publishing Company, Dordrecht, p. 151, 1980
200. Fabry, M. E., Koenig, S. H., Schillinger, W. E.: J. Biol. Chem. 4256 (1970)
201. Lanir, A., Gradstajn, S., Navon, G., FEBS Lett. *30*, 351 (1973)
202. Bertini, I. et al.: Inorg. Chim. Acta *23*, L15 (1977)

203. Bertini, I. et al.: Biochem. Biophys. Res. Commun. *78*, 158 (1977)
204. Bertini, I., Luchinat, C., Scozzafava, A.: FEBS Lett. *87*, 92 (1978)
205. Bertini, I. et al.: J. C. S. Dalton 1269 (1978)
206. Bertini, I. et al.: Inorg. Chim. Acta *36*, 9 (1979)
207. Jacob, G. S., Brown, R. D., Koenig, S. H.: Biochem. Biophys. Res. Commun. *82*, 203 (1978)
208. Wells, J. W., Kandel, S. I., Koenig, S. H.: Biochemistry *10*, 1989 (1979)
209. Fabry, M. E.: J. Biol. Chem. *253*, 3568 (1978)
210. Koenig, S. H., Brown, R. D., Jacob, G. S.: Biophysics and Physiology of Carbon Dioxide (ed. Bauer, C., Gros, G., Bartels, H.), Springer, Berlin, p. 238 (1980)
211. Koenig, S. H., Brown, R. D.: Ann. N. Y. Acad. Sci. *222*, 752 (1973)
212. Unpublished results from this laboratory
213. Solomon, I.: Phys. Rev. *99*, 559 (1955)
214. Bloembergen, N.: J. Chem. Phys. *27*, 572 (1957)
215. Bloembergen, N., Morgan, L. O.: J. Chem. Phys. *34*, 842 (1961)
216. Koenig, S. H.: J. Magn. Res. *31*, 1 (1978)
217. Bertini, I.: In: ESR and NMR of Paramagnetic Species in Biological and Related Systems (ed. Bertini, I., Drago, R. S.), NATO-ASI D. Reidel Publishing Company, Dordrecht, p. 201 (1980)
218. Hunt, H. R., Taube, H.: J. Am. Chem. Soc. *80*, 2642 (1958)
219. Meier, P. et al.: J. C. S. Chem. Comm. *36* (1977)
220. Bloch, J., Navon, G.: J. Inorg. Nucl. Chem. *42*, 693 (1980)
221. Luz, Z., Shulman, R. G.: J. Chem. Phys. *43*, 3750 (1965)
222. Rose, K. D., Bryant, R. G.: J. Am. Chem. Soc. *102*, 21 (1980)
223. Coleman, J. E.: Biochemistry *4*, 2644 (1965)
224. Barzi, D. et al.: Inorg. Chim. Acta *36*, L431 (1979)
225. Bertini, I., Luchinat, C., Scozzafava, A.: Inorg. Chim. Acta *46*, 85 (1980)
226. Bertini, I., Luchinat, C., Scozzafava, A.: In: Biophysics and Physiology of Carbon Dioxide (ed. Bauer, C., Gros, G., Bartels, H.), Springer, Berlin, p. 151, 1980
227. Lindskog, S., Ehrenberg, A.: J. Mol. Biol. *24*, 133 (1967)
228. Aasa, R., Hanson, M., Lindskog, S.: Biochim. Biophys. Acta *453*, 211 (1976)
229. Carlin, R. L.: In: Transition Metal Chemistry (ed. Carlin, R. L.), Marcel Dekker, Inc., New York, p. 1, 1966
230. Rosenberg, R. C., Root, C. A., Gray, H. B.: J. Am. Chem. Soc. *97*, 21 (1976)
231. Morassi, R., Bertini, I., Sacconi, L.: Coordin. Chem. Rev. *11*, 343 (1973)
232. Bertini, I., Gatteschi, D., Scozzafava, A.: Inorg. Chim. Acta *13*, 145 (1975)
233. Bertini, I. et al.: Inorg. Chem. *14*, 1639 (1975)
234. Bertini, I., Gatteschi, D., Scozzafava, A.: Inorg. Chem. *14*, 812 (1975)
235. Jorgensen, C. K.: Modern Aspects of Ligand Field Theory, North Holland Publishing Co., Amsterdam (1971)
236. Bertini, I., Luchinat, C., Scozzafava, A.: Inorg. Chim. Acta *22*, L23 (1977)
237. Bergstén, P. C. et al. In: Proceedings of the Alfred Benzon Symp. IV (ed. Rorth, M., Astrup, P., Munksgaard), Copenhagen, p. 363 (1972)
238. Bertini, I. et al.: Inorg. Chim. Acta *46*, 211 (1980)
239. Coleman, J. E., Coleman, R. V.: J. Biol. Chem. *247*, 4718 (1972)
240. Holmquist, B., Kaden, T. A., Vallee, B. L.: Biochemistry *14*, 1454 (1975)
241. Grell, E., Bray, R. C.: Biochim. Biophys. Acta *236*, 503 (1971)
242. Kennedy, F. S. et al.: Biochem. Biophys. Res. Commun. *48*, 1533 (1972)
243. Desideri, A. et al.: Biophys. Chem. *8*, 267 (1978)
244. Haffner, P. H., Coleman, J. E.: J. Biol. Chem. *248*, 6630 (1973)
245. Taylor, J. S., Coleman, J. E.: J. Biol. Chem. *246*, 7058 (1971)
246. Cockle, S. A.: Biochem. J. *137*, 587 (1974)
247. Haffner, P. H., Coleman, J. E.: J. Biol. Chem. *250*, 996 (1975)
248. Bencini, A. et al.: J. Inorg. Biochem. (1980) *14*, 81 (1981).
249. Bencini, A., Gatteschi, D.: Transition Metal Chemistry, in the press
250. Bencini, A., Gatteschi, D.: Inorg. Chem. *18*, 2137 (1979)
251. Horrocks, W. DeW, Jr., Burlone, D. A.: Inorg. Chim. Acta *35*, 165 (1979)
252. Bencini, A. et al.: Inorg. Chem. *18*, 2137 (1979)

253. Guggenberger, L. J. et al.: Inorg. Chem. *12,* 508 (1073)
254. Bencini, A. et al: Inorg. Chem. *18*, 2526 (1979)
255. Bencini, A. et al: Inorg. Chem. *19,* 3027 (1980)
256. Bertini, I. et al.: Inorg. Chim. Acta *46*, L91 (1980)
257. Bertini, I. et al.: Inorg. Chem. *20,* 1670 (1981)
258. Taylor, J. S., Coleman, J. E.: J. Biol. Chem. *248*, 749 (1973)
259. Bertini, I., Scozzafava, A.: In: Metal Ions in Biological Systems (ed. Sigel, H.) Marcell Dekker, New York, vol XII in the press
260. Bertini, I., Borghi, E., Luchinat, C.: J. Am. Chem. Soc. *101*, 7069 (1979)
261. Unpublished results from this laboratory
262. Hathaway, B. J., Billing, D. E., Coordin. Chem. Rev. *5*, 143 (1969)
263. Bencini, A. et al.: Inorg. Chem. *17*, 3194 (1979)
264. Haffner, P. H., Coleman, J. E.: J. Biol. Chem. *248*, 6626 (1973)
265. Bertini, I. et al.: J. Inorg. Biochem., in the press
266. Jørgensen, C. K., "Absorption Spectra and Chemical Bonding in Complexes", Pergamon Press, New York (1962)
267. Sacconi, L.: Trans. Metal Chem. *4*, 199 (1968)
268. Meirovitch, E., Lanir, A.: Chem. Phys. Lett. *53*, 530 (1978)
269. Haffner, P. H., Goodsaid, F., Coleman, J. E.: J. Biol. Chem. *249*, 6693 (1974)
270. Shinar, H., Navon, G.: Eur. J. Biochem. *93*, 313 (1979)
271. Meares, C. F. et al.: Proc. Natl. Acad. Sci. USA *64*, 1155 (1969)
272. Armitage, I. M. et al.: J. Magn. Res. *29*, 375 (1978)
273. Coleman, J. E.: In: Biophysics and Physiology of Carbon Dioxide (ed. Bauer, C., Gros, G., Bartels, H.), Springer, Berlin 1980 p 133
274. Jonsson, N. B. H. et al.: Proc. Natl. Acad. Sci. USA *77*, 3269 (1980)
275. Bertini, I., Luchinat, C., Scozzafava, A.: Biochim. Biophys. Acta *452*, 239 (1976)
276. Bertini, I., Luchinat, C., Scozzafava, A.: J. C. S. Dalton 1962 (1977)
277. Bertini, I., Luchinat, C., Scozzafava, A.: Bioinorg. Chem. *7*, 225 (1977)
278. Bertini, I. et al.: J. Inorg. Biochem. *11*, 49 (1979)
279. Wolpert, H. R., Strader, C. D., Khalifah, R. G.: Biochemistry *16*, 5717 (1977)
280. Alberti, G. et al.: Biochem. Biophys. Acta, *16,* 688 (1981)
281. Yeagle, P. L., Lochmüller, C. H., Henkens, R. W.: Proc. Natl. Acad. Sci. USA *72*, 454 (1975)
282. Stein, P. J., Merrill, S. H., Henkens, R. W.: J. Am. Chem. Soc. *99*, 3194 (1977)
283. Patterson, A., Ettinger R.: Z. Elektrochemie *64*, 98 (1960)
284. Riddiford, L. M.: J. Biol. Chem. *239*, 1079 (1964)
285. Riddiford, L. M. et al.: J. Biol. Chem. *240*, 3305 (1965)
286. Nilsson, A., Lindskog, S.: Eur. J. Biochem. *2*, 309 (1967)
287. Clementi, E. et al.: Montedison Tech. Rep. DDC *802*, December 1978
288. Clementi, E. et al.: FEBS Lett. *100*, 313 (1979)
289. Clementi, E. et al.: Gazz. Chim. Ital. *109*, 669 (1979)
290. Clementi, E., Scordamaglia, R., Cavallone, F.: J. Am. Chem. Soc. *99*, 5545 (1977)
291. Clementi, E., Bolis, G.: J. Am. Chem. Soc. *99*, 5550 (1977)
292. Bertini, I., Luchinat, C., Scozzafava, A.: FEBS Lett. *93*, 251 (1978)

K. Notes Added in Proofs

1) Recent ^{17}O relaxation studies on the copper derivative have shown that oxygen is bound to the metal in the pH range investigated (Bertini, I., Canti, G., Luchinat, C.: Inorg. Chim. Acta *56,* 1 (1981)) and that it exchanges rapidly on the NMR time scale. It is suggested that if OH^- is present, it might exchange through an equilibrium of the sort

$$M - {}^{*}OH + H_2O \rightleftharpoons M - {}^{*}OH(H_2O) \rightleftharpoons M - OH + H_2O^{*}$$

2) Recent water ^{1}H T_1 studies at various magnetic fields have allowed to establish that the derivatives assigned as five coordinated have a water molecule in the coordination sphere and τ_c as low as 10^{-12} s, whereas those derivatives which have been assigned as pseudotetrahedral do not contain bound water and τ_c is ca. 10^{-11} s (Bertini, I., Canti, G., Luchinat, C.: Inorg. Chim. Acta *56* (1981)).
3) ^{1}H NMR studies of the cobalt protein have shown that the three histidine residues remain coordinated upon binding of inhibitors (Bertini, I., Canti, G., Luchinat, C.: J. Am. Chem. Soc, in press).
4) An alternative suggestion has been made by S. H. Koenig et al. (Jacobs, G. S., Brown, III, R. D., Koenig, S. H.: Biochemistry *19,* 3754 (1980)) who proposed aniline to bind to the coordinated water. This hypothesis, however, does not account for the sizeable ^{13}C contact shifts measured for the coordinated ligand[169]. In the same paper the methyl proton relaxation rate of CH_3OD in presence of CoCA has been reported to be constant with pH.

Scope and Limitation of Single Electron Transfer in Biology*

Peter Hemmerich, Vincent Massey, Heinrich Michel and Christian Schug**

Fakultät für Biologie, Universität Konstanz, 7750 Konstanz, Germany

This article is thought to be a supplement to an earlier one "Mechanism of Hydrogen Transfer in Redox Enzymes"[33]. Most of the basic concepts and hypotheses put forward then, are still valid. Substantial new evidence supporting our views has been advanced since then. The new data presented here concern modified or as we call them "mutilated flavins", namely 5-deaza- and 5-thiaflavin, which upon incorporation into apoflavoprotein retain some, but not all of the activities encountered with the native flavoproteins.

---

* Presented at the International Symposium "Single Electron Transfer in Biology", held in Konstanz from 9th–13th March 1981, in honor in Prof. Dr. Helmut Beinert

** Permanent guest professor at the University of Konstanz. Permanent address: Department of Biological Chemistry, University of Michigan, Ann Arbor, Michigan 48109, USA

Introduction

In the beginning of biochemical experimental work on the mechanism of biological oxidation, there was great uncertainty about the nature of biological redox carriers: It was, however, obvious that redox equivalents must be "carried" through many steps, starting e.g. from alcoholic substrates in the citric acid cycle, and ending with the final acceptor, molecular oxygen. In the early 30's the founding fathers of redox enzymology, such as Heinrich Wieland, began to subdivide respiration into a cyanide sensitive and a cyanide insensitive part[89)]. By this, they came close to the idea that cyanide insensitive respiration was connected with hydrogen, and cyanide sensitive with oxygen transfer. Single electrons and radicals were at this time believed to be non-existent as chemical entities.

Even nowadays, loss with the assignment of specific redox centers to cyanide sensitive and insensitive respiration is somewhat equivocal, especially with the non-metallic redox centers, such as reduced flavin. As will be discussed in more detail below, flavin-dependent O_2-activation can, in principle, follow $1\,e^-$ stoichiometry (with production of superoxide) or $2\,e^-$ stoichiometry (with direct production of peroxide). We have little information as to what extent these centers are naturally accessible to oxygen. Cyanide insensitive respiration does not conserve energy. It is still an open question, whether this is an unsuccessful pathway, perhaps due to the inefficiency of nature to preclude O_2-contact with reduced centers of high O_2-affinity, or whether nature requires a certain level of superoxide formation in tissues for anabolic purposes, which in turn is regulated by the detoxifying action of catalase and dismutase. In any case, cyanide insensitive respiration must be differentiated according to its stoichiometry: Iron-sulfur centers, of course, transfer single electrons only, while flavin centers are ambiguous.

Lenor Michaelis then came with his discovery of biological radicals, namely flavosemiquinones, emerging from his qualitatively perfect, though quantitatively incorrect electrochemical evaluation of the flavoquinone redox system[67)] (cf. Eq. 1).

$$Fl_{ox} \underset{}{\overset{e^-/H^+}{\rightleftharpoons}} H\dot{F}l \overset{e^-/H^+}{\rightleftharpoons} H_2Fl_{red} \tag{1}$$

At the same time Richard Kuhn reached a goal, which was the dream of a chemist, a crystalline preparation of (flavosemiquinone) radical[50)]. The two types of radicals he prepared in solid state were, however, of entirely artificial nature, the one being the red semiquinone cation, the acidity of which (pK ~ 2) precludes biological relevance, and the other one a dimeric (flavoquinhydrone) complex, which is largely in the disproportionated form. Though in principle, flavin/flavin contact or interaction could be biologically relevant[32)], it should be pointed out that such a flavin sandwich conformation has not been found up to date in any protein. Furthermore, Richard Kuhn anticipated, without discussion, a 1-protonated structure for the neutral semiquinone[50)] (cf. Fig. 14). This seemingly small mistake, nevertheless, precludes an understanding of the operational behaviour of the radical state in flavoproteins, as will be shown further on.

The discovery of the fact that organic compounds containing single electrons might be absolutely stable and crystallizable, led Michaelis to the great simplification that in biological redox reactions no other valence changes than of ± 1 should be permitted[68)].

Thus, it was anticipated that even in those reactions, where no radicals could be observed, they were still present as intermediates, but would decay too rapidly for analytic inspection. Nowadays, the maintainance of this postulate would limit rates of "electron pairing" down at the picosecond level[45].

2 e^--Transfer

It took more than 30 years until the Michaelis hypothesis could be definitely eliminated. Still the very best example for a biological redox reaction of non-radical character is nicotinamide-dependent (de)hydrogenation. Very good evidence for the extreme destabilization of the radical level in the 2 e^--redox system $NAD^+/NADH$ has been put forward by the work of Blankenhorn[3]. He compared nicotinamide models with the sterically flavin-shaped analog 5-deazaflavin and found that the change in the steric shape did not change at all the electronic behaviour. As Table 1 shows, nicotinamide, (and respectively 5-deazaflavin), requires for the uptake of the first single electron a potential of −850 (−770) mV, while the second electron will follow with a much higher potential of +210 (+130) mV. The 2 e^--redox couples $NAD^+/NADH$, as well as (dihydro)-deazaflavin are in the range of −320 mV[38]. Hence, these redox chromophores strictly prefer uptake and donation of 2 e^--equivalents over the reactions with single electrons (cf.

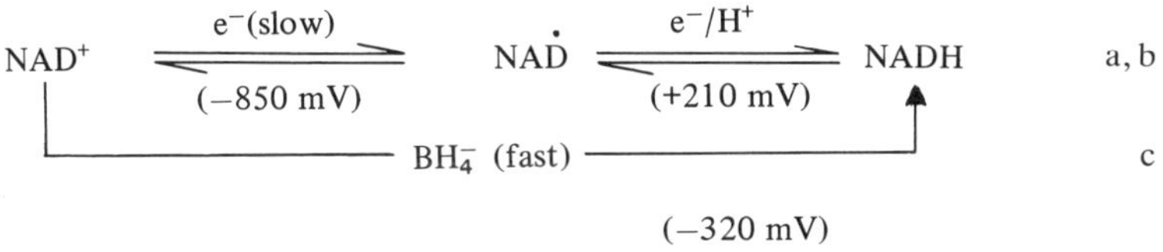

(2)

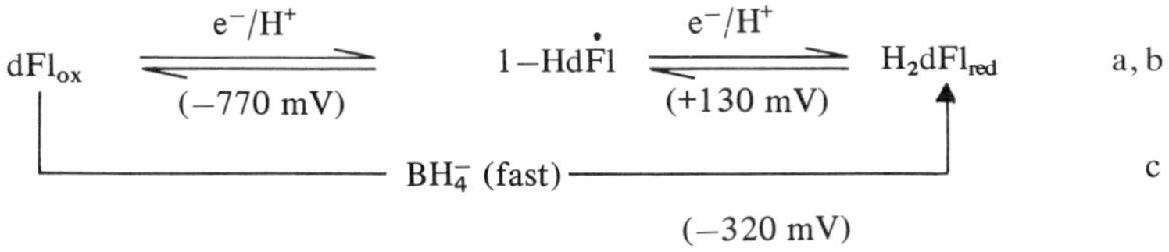

(3)

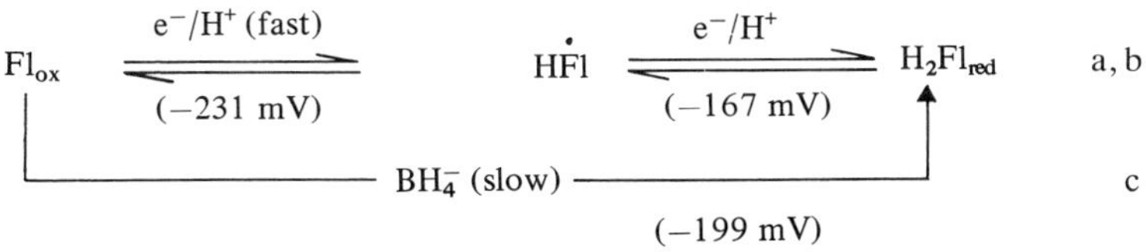

(4)

Table 1. Enzyme and model redox potentials E_0' (pH 7) (mV)[a]

System	$Fl_{ox} \leftrightharpoons Fl^{\bullet}$ E_2 (1 e^-)	$Fl_{ox} \leftrightharpoons Fl_{red}$ E (2 e^-)	$Fl^{\bullet} \leftrightharpoons Fl_{red}$ E_1 (1 e^-)	$\Delta E_{1,2}$
Free Fl (1 e^- or 2 e^-)	−231	−199	−167	−64
1-RFl (2 e^--only)[b]	not obtained	−009	not obtained	
5-RFl (1 e^--only)[b]	+423[e]	+216[c]	+010	+413
Flavodoxins (1 e^--only)	−190 to −130	−280[c]	−420 to −377	+240
NADP-reductase (1 e^- or 2 e^-)	−320	−360	−400	+80
Glucose Oxidase (2 e^--only) pH 5.3	−046	−051	−056	+10
Glucose Oxidase (2 e^--only) pH 9.3	−200	−220	−240	+40
5-Thiaflavin (1 e^--only)	irrev.	irrev.	+430	
5-Deazaflavin (2 e^--only)	−770	−320	+130	−900
Nicotinamide (2 e^--only)	$NAD^+ \leftrightharpoons NAD^{\bullet}$ −850	$NAD^+ \leftrightharpoons NADH$ −320	$NAD^{\bullet} \leftrightharpoons NADH$ +210	−1060
Dithionite[d] (1 e^- or 2 e^-)	$SO_2 \overset{1e^-}{\leftrightharpoons} \dot{SO_2^-}$ −660	$2\,SO_2 \overset{2e^-}{\leftrightharpoons} (SO_2)_2^{2-}$ −386		
Oxygen[d] (1 e^-, 2 e^- or 4 e^-)	$O_2 \overset{1e^-}{\leftrightharpoons} \dot{O_2^-}$ −330	$O_2 \overset{2e^-}{\leftrightharpoons} H_2O_2$ +290	$O_2 \overset{4e^-}{\leftrightharpoons} H_2O$ +820	

[a] Data of Ref. 37 completed
[b] 1-RFl_{ox} is at pH 7 present as N(3) deprotonated mesoionic neutral species, whereas 5-RFl_{ox} forms the pseudobase 5-RFl-4a-OH.
[c] calc. as arithm. means of $E_{1,2}$
[d] given for comparison
[e] Extrapolated from pH 4 for the cation 5-RFl_{ox}^+ (cf. Fig. 8 b)

Eqs. 2 and 3). In contrast to this are the redox properties of flavins, which show slow reaction with a hydride donor, but fast reaction for the 1 e^- uptake (cf. Eq. 4).

The Uniqueness of Nicotinamide

Nevertheless, one can force single electrons into the nicotinamide or deazaflavin system by *photo*chemical means, using the same 5-deazaflavin as photosensitizer and an organic 1 e^--photoelectron donor such as EDTA as substrate[60]. Table 2 reviews the events

Table 2. Transfer mode and reaction stoichiometry of chemical and photochemical reductions of flavin (upper half) and nicotinamide (lower half). Note that the oxidants (Fl_{ox} and NAD^+) as well as the reductants (substrates) have their separate influence on redox-transfer mode stoichiometry, namely, either $1\,e^-$ or $2\,e^-$ (i.e. H^- or R^-) or ambiguous. Smooth reactions can in general only occur when transfer mode of donor and acceptor are the same, or if at least one of the reactants has an ambiguous character. Whenever the ambiguous reactions are in principle reversible the system lends itself as a $1\,e^-/2\,e^-$ interconverting redox catalyst. The very slow dark reaction between Fl_{ox} and BH_4^- proves that the ground-state flavin is no H^-- but instead an R^--carrier. Note further that $2\,e^-$-transfer can not occur without the help of a $2\,e^-$-carrier, namely, either R or H, i.e. "double $1\,e^-$-transfer", rapid subsequent uptake of two single electrons has never been verified in any model reactions

Acceptor (oxidant)	Transfer-mode and stoichiometry postulated	observed	Substrate (reductant)		Transfer-mode and stoichiometry of substrate	(primary) product(s)	Secondary reaction	Ref.
Fl_{ox}	ambiguous	$1\,e^-$	PhOH	$h\nu$	$1\,e^-$ (obs.)	$H\dot{F}l + Ph\dot{O}$[a]	back donation	80)
			EDTA	$h\nu$		$H\dot{F}l$[b]	dismutation	22)
			dithionite[c]	dark		$H\dot{F}l + SO_2$	dismutation	51)
			H_2Fl_{red}	dark $h\nu$		$Fl_{ox} + H_2Fl_{red} \underset{h\nu}{\overset{dark}{\leftrightharpoons}} 2\,H\dot{F}l$		18) 40)
		H^-	BH_4^-	dark	H^- (obs.)	slow reduction to H_2Fl_{red}		71)
				$h\nu$		H_2Fl_{red}[d]		40)
		ambiguous	Ph–CHOH–COO^-	$h\nu$	ambiguous	H_2Fl_{red} + Ph–CO–COO^-		85)
		R^-	Ph–CH_2–COO^-	$h\nu$	R^- (obs.)	4 a- *and* 5-R,$HFl_{red} + CO_2$		84)
			NADH	dark	R^- (postulated)	NAD-FlH	$H_2Fl_{red} + NAD^+$	this lecture
NAD^+	H^-, R^-	$1\,e^-$ (enforced)	$1\,e^-$ aq[e]	dark	$1\,e^-$ (obs.)	$NAD^{\cdot}$	dimerisation	9)
		H^-	BH_4^-	dark	H^- (obs.)	NADH		42)
		R^-	CN^-	dark	R^- (obs.)	NAD-(4)-CN		86)
			dithionite	dark[f]		NAD-(4)-SO_2^-	NADH + SO_2	4)

[a] PhOH = phenol, back donation yields Fl_{ox} + PhOH; [b] Decay of the substrate radical $EDTA^\bullet$ is being investigated in our laboratory; [c] Reducing agent: $SO_2^{\bullet -}$; [d] Under continuous illumination addition of H^- at carbonyl C(4) is observed[71)]; [e] From pulse radiolysis; [f] Reducing agent: SO_2^{2-}, proposed decay of the primary adduct: NAD-(4)-SO_2^- $\rightarrow$ $NAD^+ + HSO_2^-$ $\rightarrow$ NADH + SO_2

which can be observed in such photoreductions, depending on the nature of the substrate. Starting with cyanide or borohydride as extreme cases, we find even under photochemical conditions a strict $2e^-$-transfer consisting of nucleophilic addition of negatively charged substrate residue to the excited chromophore. Analogous dark reactions are rare in chemical systems (cf. Eq. 4), but are found in the dark reaction between flavin and reduced nicotinamide, which is, of course, too slow to have any biological relevance[77]. It is, therefore, the aim of $2e^-$-redox proteins to circumvent the chemical difficulty of providing $2e^-$-equivalents for oxidoreductive purposes. The system $NAD^+/NADH$ (as shown in Fig. 1) is by nature uniquely provided to fulfill this aim, since the oxidized coenzyme can not easily accept single electrons for chemical reasons. On the other hand, the reduced coenzyme is ready to split out hydride equivalents in order to reach a state of aromatization. Removal of a proton from NADH would result in the formation of the anti-aromatic ion NAD^- at high expense of energy[19]. Figure 2 shows examples of anti-aromatic species which are chemically relevant. All of them are high energy intermediates, which undergo immediate stabilization by rearrangement. In order to transfer

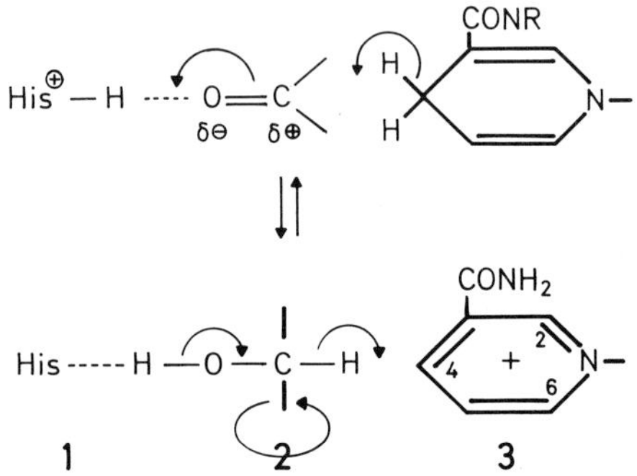

Fig. 1. Mechanism of nicotinamide dependent hydride transfer[1]. Note that the apoprotein needs not only activate position 4 but also protect positions 2 and 6. This is the reason why there are no simple chemical models for NAD(H)

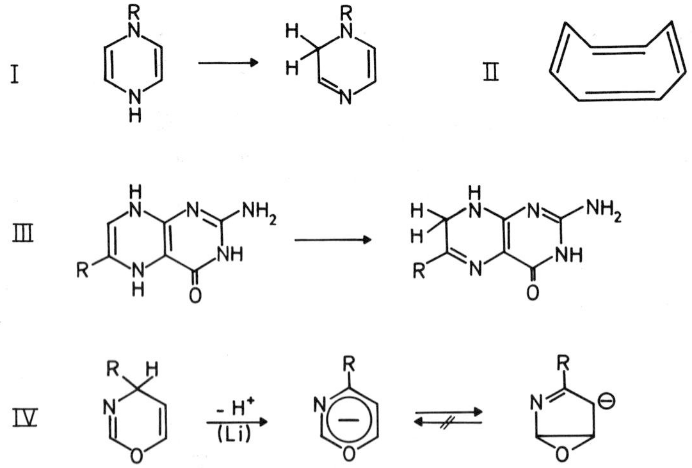

Fig. 2. Examples for potentially antiaromatic species, which avoid this energy rich state either by prototropic shift (I, III)[12, 13, 74] or by deplanarization (II), or by valence isomerization (IV)[75]

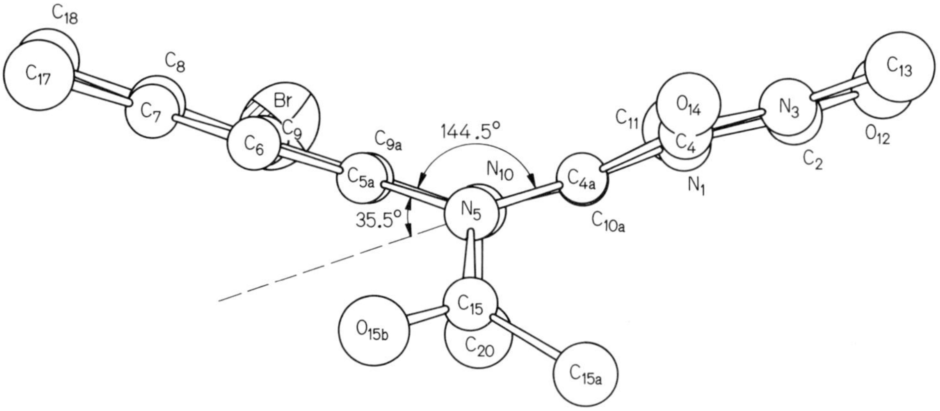

Fig. 3. Bent (crystal) structure of (5-acetyl)-dihydroflavin[47]. Flattening of the molecule leads to antiaromaticity. Note that the N(5)-acetamide group is also nonplanar in itself and is easily hydrolysable because of lacking amide resonance[44]

hydride equivalents, nature has found one of the most simple chromophores, NAD^+, to fulfill this requirement. Another biologically relevant case of anti-aromaticity is also shown in Fig. 3, namely dihydroflavin. Since this more delocalized π-electron system has no CH-bond, from which a hydride equivalent could be split, but only NH-bonds (which do not lend themselves as easily for rupture because of electronegativity), the dihydroflavin system escapes from anti-aromaticity simply by deplanarization, which precludes tricyclic delocalization of π-electrons. This folded structure of dihydroflavin is of great biological importance, as will be shown below. The planar conformation is high in energy, anti-aromaticity implying as in all anti-aromatic systems, a low lying triplet state[11]. The low lying triplet is a prerequisite for the tendency of the chromophore to accept and donate single electrons. In contrast to nicotinamide, flavin is thus not a hydride donor-acceptor, which is demonstrated by the fact, that in contrast to nicotinamide, flavin reacts with borohydride only extremely slowly[71] (cf. Eq. 4c).

Finally, when high energy single electrons are forced into a $2e^-$-specific acceptor system, e.g. by the above mentioned photochemical method of Massey and Hemmerich[60], the resulting radicals such as $NAD^\bullet$ will undergo dimerization with subsequent redox-inactivation of the coenzyme[9] (cf. Eq. 5).

$$NAD^+ \xrightarrow{\text{EDTA, h}\nu} NAD^\bullet \rightarrow \tfrac{1}{2}(NAD)_2 \quad (5)$$

In spite of the tremendous progress that has been made since the early 30's, it is very curious to see that our current textbooks still struggle with the same difficulties. Even worse, many authors are not aware of this and confine themselves to the description of ever more sophisticated redox systems and redox chains, which then turn out to be redox networks, while at the same time, they can not adopt non-equivocal nomenclature, concerning the nature of redox transport and redox carriers. Table 3 shows an evaluation of current textbooks in this respect, and demonstrates the state of present knowledge.

Table 3. Mechanism of redox transfer mode stoichiometry. While the alcohol/nicotinamide hydride transfer is now widely acknowledged, the flavin versatility as intended by nature remains largely unclear. Furthermore, the change in transfer-mode between nicotinamide and flavin ($H^- \rightarrow R^-$) remains without discussion. Hence, the flavin is assumed to be at will a hydride transfer cofactor and the uniqueness of nicotinamide is lost

	Transfer mechanism supposed for		
Textbook	Alcohol/NAD^+	NADH/flavin	Other flavin substrates/flavin
Mahler and Cordes[55]	H^-	?	1 e^-
Karlson[46]	H^-	?	1 e^-
Buddecke[8]	H^-	?	?
Lehninger[53]	H^-	?	1 e^-, 2 e^-
Metzler[66]	H^-	?	1 e^-, 2 e^-, H^-, R^-
Dixon and Webb[14]	H^-	?	?
Stryer[78]	H^-	?	?
Walsh[87]	H^-	H^-	1 e^-, R^-
this lecture	H^-	R^-	1 e^-, R^-, H^-

? Transfer mechanisms are not specified

Biological Redox Packages

Firstly, we want here to differentiate between different types of redox packages, those which are unimportant in biology, others, which remain questionable, and a third category, which is biologically significant, although such available evidence is largely neglected, as Table 3 shows clearly.

Hydrogen Packages. Starting with *molecular hydrogen,* we have a redox package, which can only be digested biologically by the iron sulfur protein hydrogenases. The term hydrogenase is already extremely misleading, since it is synonymous with reversal of dehydrogenase. The mode of dihydrogen splitting in hydrogenase is strictly 2 e^- and thus excludes the occurrence of intermediate $H^\bullet$-packages, in the course of H_2-activation[81]. The hydrogen atom is a well known species in physical chemistry but contrary to common opinion, hydrogen atoms are kinetically rather stable[15], compared with rates of biocatalysis, for several reasons:

(a) The hydrogen atom (in the gaseous state) is a species of relatively high diameter (1.06 Å) which is highly symmetric and requires polarization for activation.

(b) In polar solution the hydrogen atom is an acid of approximate pK 9[30]. The pK is not too well defined, because as mentioned before, polarization is required for further reaction, and therefore, the deprotonation of the hydrogen atom is not a very fast reaction. But in spite of this we can at least state, that at physiological pH and in hydrophilic environment we have to account for hydrogen atoms only in their deprotonated forms, namely hydrated electrons, which are kinetically much more labile.

Hence, hydrogen atoms do not lend themselves as biocatalytic redox packages. The only hydrogen package to be considered in this context would thus be hydride.

The hydride ion as found in sodium hydride, however behaves chemically as a base and not primarily as a reductant (cf. Eq. 6).

$$CH_3\text{–}CO\text{–}CH_3 + NaH \rightarrow CH_3\text{–}CO\text{–}CH_2^- Na^+ + H_2 \tag{6}$$

If applied to a keto-compound, such as acetone, hydride will react preferably by abstraction of an α-proton rather than by reduction of the carbonyl group. This can only be overcome by an exchange of "simple" hydride towards complex hydride, such as BH_4^-, which lacks the symmetry of the H^--ion[7] (cf. Eq. 7).

$$CH_3\text{–}CO\text{–}CH_3 \quad + \quad BH_4^- \xrightarrow{H_2O} CH_3\text{–}CHOH\text{–}CH_3 + B(OH)_3 \tag{7}$$

Thus, in redox enzymology, hydride can not occur as a chemical entity of whatever short lifetime; instead, hydride in biology is confined to be a transition state, the occurrence of which is uniquely connected with a very special coenzyme, nicotinamide, as already mentioned and shown in Figs. 1 and 4.

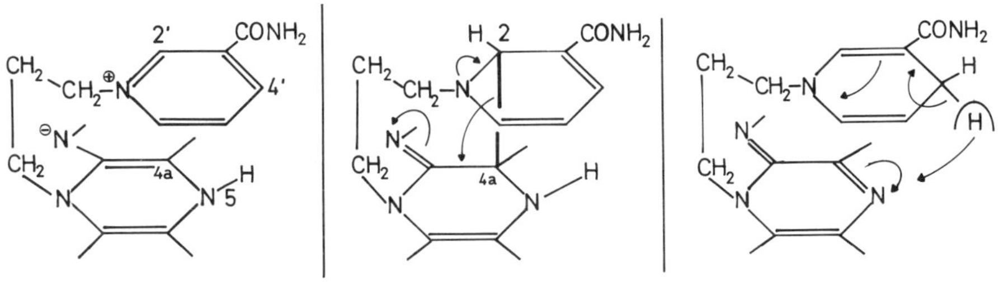

Fig. 4. Hypothetical $2\,e^-(\sigma)$-transfer between flavin and nicotinamide: A covalent bond is formed and split between the heterocyclic nuclei with subsequent proton transfer. Note that a steric correlation between the σ-bonding sites is required which is more specific than simple donor-acceptor π-overlap. The latter is possible but not essential for the starting and the final state. The covalent intermediate must be anticipated to be rather colorless as the formula shows and, therefore, a shortlived species difficult to detect experimentally

Oxygen Packages. Contrary to the activation of molecular hydrogen which is specifically confined to hydrogenases, nature has at its disposal quite a few systems which can achieve the difficult task to activate *molecular oxygen*, namely, various iron (heme or non-heme) proteins[10]) copper[56]), and the purely organic flavin and pteridine systems[39, 59]).

The various modes of oxygen activation are differentiated in the following equations

dioxygen reduction: $O_2 \xrightarrow{4\,e^-/4\,H^+} 2\,H_2O \quad (4\,e^-)$ [20])(8)

monooxygen insertion (oxygenation): $O_2 \xrightarrow{2\,e^-/2\,H^+/RH} H_2O + ROH\,(2\,e^- + 2\,e^-)$ [59])(9)

$$\text{Oxidase action: } O_2 \xrightarrow{2e^-/2H^+} H_2O_2 \quad (2e^-) \qquad [6](10)$$

$$\text{single electron transfer: } O_2 \xrightarrow{e^-} O_2^{\bullet -} \quad (1e^-) \qquad [64](11)$$

Discussion of the stable "dioxygen packages", namely H_2O_2 and superoxide, is out of scope in this article. Among the unstable potentially relevant "monooxygen packages", the *hydroxyl radical* should be taken with great suspicion wherever it is mentioned: The reactivity of $OH^{\bullet}$ with CH-bonds is so high, kinetically as well as thermodynamically[5], that in order to maintain $OH^{\bullet}$ as an oxygen package, nature would have to protect it against all contact with CH-bonds. A CH-free environment, however, cannot be accepted as biological. The discussion of "$OH^{\bullet}$ in biology" is therefore limited to metal complexes, acting as stabilizers of the OH-radical[5]. Since, however, metal chelation removes the characteristic features of $OH^{\bullet}$ it is a matter of semantics to postulate its occurrence in metal complexes. In any case, there is no Fe–OH bond known, which would ever dissociate homolytically under biological conditions.

For insertion of oxygen into CH-bonds, nature has, therefore, developed an "oxygenation" apparatus, working with atomic oxygen ("oxene" packages), as transition state[59,61] (cf. Eq. 9).

After this general discussion of biological redox packages, we can now come back to the initial subject of this review, namely scope and limitation of single electrons as the "most simple" redox packages in biology. The key question in this context: Where in the natural redox chains is the borderline between dehydrogenation and single electron transport or even more precisely, when are electrons disconnected from organic residues and "mobilized"? Is the locus of this disconnection identical with the site of "pair splitting"?

We have already discussed that the CH-substrates of nicotinamide-dependent enzymes must be preactivated in the sense of C^+H^-. It turns out that the reverse is true for the substrates of flavin-dependent dehydrogenases, which must be preactivated in the sense C^-H^+[36]. The resulting carbanion R^- is, of course, transferred not as a chemical entity, but in a transition state, and may yield a flavin covalent adduct, possible structures of which are shown in Fig. 5.

All biological redox chains of mitochondrial, microsomal or chloroplast origin, invariably contain one and only one critical point of a working subunit which converts carbanionic residues R^- into two single electrons and a carbocation R^+ and vice versa (R^+ being further stabilized by hydrolysis).

The thus defined "electron pairing" (or electron pair splitting) implies two problems, which are only at first sight easy to resolve, namely a charge correlation and a spin correlation problem.

The charge correlation invokes the steady rapid disposal of a positive carrier unit (the already mentioned R^+), which neutralizes the repelling power between the two electrons. The spin correlation invokes the steady presence of a spin relaxing, i.e. spin-orbit coupling heavy atom center, particularly at least one sulfur iron cluster, which at the same time serves as a single electron store (see below). By this fact, the flavin-dependent $2e^-$-splitting differs from any $1e^-$/$2e^-$-transformation, which might occur in the lipoquinone pools of redox chains[43].

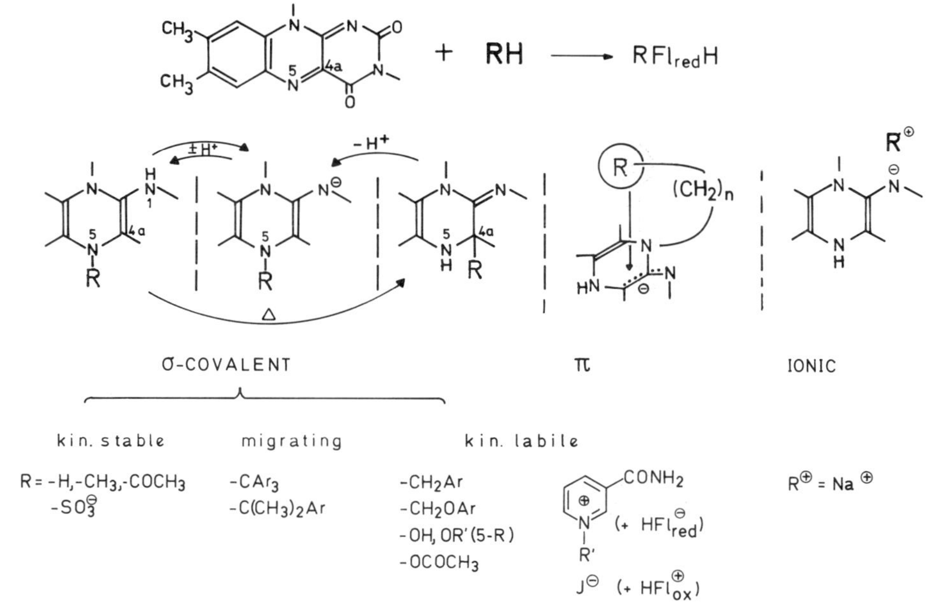

Fig. 5. Bonding modes in flavin-substrate complexes as exemplified by alkyl-dihydroflavins $RFl_{red}H$. The RFl-bond can assume a wide variety of properties: (**1**) When R^+ is more stable than R^-, such as in R = acetyl, σ-bonding is limited to N(5)[31]. (**2**) When, on the contrary, R^- is more stable than R^+, such as in R = cyclopentadienyl, stable σ-bonding is limited to C(4 a)[48]. (**3**) In the case of R = H statement 1 is verified thermodynamically[24]. (**4**) In the case of R = CH_3 kinetically stable 4 a- *and* 5-isomers can be isolated[24]. (**5**) When the substrate model residues R become more labile, such as R = $ArC(CH_3)_2$, pH-regulated sigmatropic migration and backmigration is observed[35] (cf. Eq. 12). (**6**) At higher temperatures migrations are accelerated to yield fluctuations[35]. A change from σ- to π-character of the RFl-bond is observed e.g. for R = iodide (+Fl_{ox} acceptor) or R = NAD^+ (+$Fl_{red}H_2$ donor)[2]. But there is a very small energy difference between σ- and π-"complexes". The extreme of an entirely ionic RFl-bond is reached for metallic R[36]

$$4\,a\text{-R-Fl}_{red}\text{H} \rightleftarrows 5\text{-R-Fl}_{red}\text{H} \quad (12)$$

The Uniqueness of Flavins

In order to understand the versatility and the uniqueness of flavin coenzymes, one must always keep in mind that flavin enzymes exhibit three mechanistically independent activities:

(A) (de)hydrogenation, implying RH-deprotonation, electron pair splitting and R^+-hydrolysis,

(B) single electron transfer (towards $1e^-$-storage and energy conserving centers or membrane areas),

(C) dioxygen activation.

Note though, that only the $1e^-$-mode of O_2-activation is within the scope of this lecture.

The three main flavin activities have to be combined pairwise by input and output. This results in the distinction of five classes of flavoproteins[41, 63], namely:

1) A $\leftrightarrows$ A: dehydrogenases (or with a better term, transhydrogenases: group transfer only) ($2e^-$ only)
2) A $\leftrightarrows$ C: dehydrogenases/oxidases ($2e^-/2e^-$) (cf. Eq. 10)
3) A $\leftrightarrows$ C: dehydrogenases/oxygenases ($2e^-/4e^-$) (cf. Eq. 9)
4) A $\leftrightarrows$ B: (de)hydrogenases/e^--transferases ($2e^-/1e^-$)
5) B $\leftrightarrows$ B: pure e^--transferases (flavodoxins: $1e^-$ only)

In contrast to lipoquinones, the spectral properties of which are less pronounced and confined to the UV-range, we can easily differentiate the biologically relevant flavin species, occurring as essential or artificial intermediates in flavin (bio)chemistry, namely Fl_{ox}, $H\dot{F}l$, $\dot{F}l^-$, HFl^-_{red} and H_2Fl_{red}. This implies one oxidized species, two (half-reduced) radicals and two fully reduced species, the physical properties of which are illustrated in Table 4.

Flavin-dependent single electron transfer, the central subject of this lecture, requires a thermodynamic stabilization of the radical state, as given by the dismutation equilibrium (cf. Eq. 13).

$$Fl_{ox} + H_2Fl_{red} \leftrightarrows 2\,H\dot{F}l \tag{13}$$

Table 4. Flavin species of biological relevance

Redox-state		protonated		neutral		deprotonated
oxidized		$(HFl^+_{ox})^a$	$\xrightarrow{pK \sim 0^{17)}}$	Fl_{ox}		
				Flavodoxin		
	λ^{abs}_{max}	390 nm[17]		447 nm[25]		
	λ^{em}_{max}	488 nm[83)b]		530 nm[25)c]		
Semiquinone		$(H_2\dot{F}l^+)^a$	$\xrightarrow{pK \sim 2.3^{52)}}$	$H\dot{F}l$	$\xrightarrow{pK \sim 8.4^{18)}}$	$\dot{F}l^-$
				Flavodoxin		Glucose oxidase, pH 10
	λ^{abs}_{max}	488 nm[17]		580, 627nm[64]		488 nm[57]
	EPR			1.9 mT[73]		1.5 mT[73)d]
reduced				H_2Fl_{red}	$\xrightarrow{pK \sim 6.3^{18)}}$	HFl^-_{red}
				Flavodoxin, pH 5		Flavodoxin, pH 8.5
	λ^{abs}_{max}			350 nm[65]		368 nm[25]
	λ^{em}_{max}			507 nm[25)e]		530 nm[25]

a These species are not found in biology
b Fluorescence emission of 2-methoxy-3-methyl-lumiflavinium-perchlorate measured in acetonitrile
c Fluorescence emission of flavoprotein is often quenched
d Electron paramagnetic resonance line width for immobilized flavoproteins, measured in mT (milli tesla), 1 mT = 10 G. Note that flavoproteins, covalently linked at C(8α) e.g. succinate dehydrogenase, exhibit a reduced line width of 1.2 mT[73]
e Fluorescence emission of reduced lactate-oxidase at pH 7.0

This equilibrium is shifted towards the left hand side under all chemical conditions (apart from pH ≤ 0, $pK_{radical} \sim 2.3$[52)]). For purposes of biological single electron transfer, this equilibrium must be regulated by the apoprotein (see below).

Hence, the biological flavins can, but need not, stabilize the radical level in the same way as they can, but need not transfer single electrons. If the apoprotein stabilizes the radical level in any case of a $2e^-$-activity, the action of the radical will be inhibitive. This is the case when the red radical $Fl^{\cdot-}$ is observed, which is in general, not a catalytically active species[23, 69)].

If flavin is indeed the natural borderline between single and double electron transfer, this leads immediately to the next question: What decides whether flavin will act in the $1\,e^-$- or $2\,e^-$-mode?

Apoprotein Regulation of Flavin Reaction Modes

An answer towards this question has been proposed recently by Hemmerich and Massey[41)]: Regulation is brought about by positive charges or hydrogen bridges from the protein to either the N(1)- or the N(5)-site of the flavin molecule, as is shown in Fig. 6. How can this be proven?

Blocking of nitrogen lone pairs by stable biological hydrogen bridges can be mimicked chemically by introduction of alkyl substituents. This is, of course, a regio-

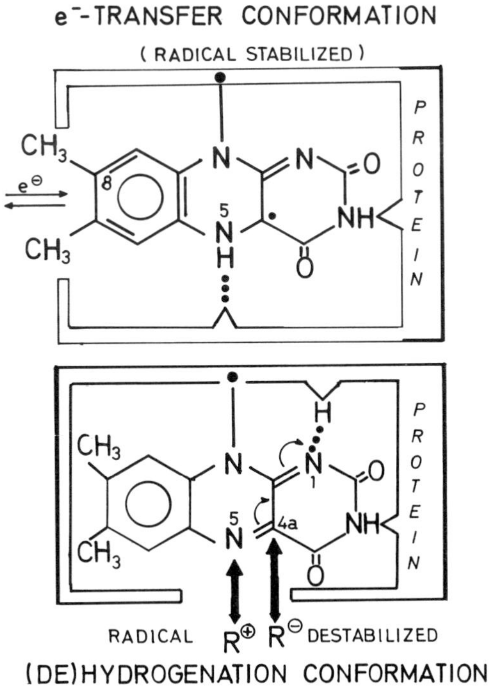

Fig. 6. $1\,e^-$/$2\,e^-$-alternative of flavin-dependent redox transfer as regulated by regio-specific hydrogen bonds or positive charges from the apoprotein towards the lone pairs of either the N(1)/O(2α)-region ($2\,e^-$-transfer, lower part) or the N(5) lone pair ($1\,e^-$-transfer, upper part). Note that in the $1\,e^-$-transfer case the input and output of electrons is presumably through the C(8)-edge while in the $2\,e^-$-transfer case the substrate (carbanion) access is through positions 4a, 5. Thus blocking and deblocking of lone pairs is accompanied by opening and closing of the access site

specific irreversible process, but it provides us with modified flavins which can be taken to be models for either specific 1 e^-- or specific 2 e^--transfer. As can be seen from Fig. 7, alkylation in position 1 or 2 α leads to a flavoquinonium system (cf. Eq. 14), which destabilizes the radical level, while, on the other hand, alkylation in position 5 has exactly the reverse effect (cf. Eq. 15).

$$1/2\,\alpha\text{-R-Fl}^+_{ox} \underset{}{\overset{2e^-/H^+}{\rightleftharpoons}} 1/2\,\alpha\text{-R-Fl}_{red}\text{H} \qquad ^{76)}(14)$$

$$\lambda_{max}\ 380\ \text{nm} \quad +136\ \text{mV} \quad \lambda_{max}\ 292\ \text{nm} \qquad \text{pH } 2.45$$

$$5\text{-R-Fl}^+_{ox} \overset{e^-}{\rightleftharpoons} 5\text{-R-Fl}^\bullet \overset{e^-/H^+}{\rightleftharpoons} 5\text{-R-Fl}_{red}\text{H} \qquad ^{76)}(15)$$

$$\lambda_{max}\ 548\ \text{nm} \quad +427\ \text{mV} \quad \lambda_{max}\ 582\ \text{nm} \quad +235\ \text{mV} \quad \lambda_{max}\ 340\ \text{nm} \qquad \text{pH } 2.45$$

In Fig. 8 we have listed the corresponding potentials and we see that in the 5-alkylated case the 1 e^--potential (cf. also Eq. 15) is elevated, while in the 1/2 α-substituted flavin (cf. also Eq. 14) the 2 e^--potential is elevated. Figure 9 shows finally, that upon reduction of 1/2 α-alkylated flavin we do not observe the slightest amount of intermediate radical, while we obtain up to quantitative formation of the blue radical upon reductive titration of N(5)-substituted flavin[76].

This is in good agreement with the fact, that the biologically active blue radical is neutral. The stable blue and active radical carries its mobile proton in position 5[73].

It is not surprising therefore, that a biocatalytically functional semiquinone has the blue neutral $HFl^\bullet$ structure and that this mobile hydrogen is kept in position by a hydrogen bridge from the protein. This is experimentally supported by the X-ray crystallographic data of flavodoxins[54].

As a further consequence, whenever a flavoprotein transforms 2 e^-- into 1 e^--packages, it must undergo each time a conformational change switching from unstable red radical i.e. "1/2 α-block", towards stable blue radical and "5-blocked", or else the required separation of 1 e^--potentials (cf. Fig. 8) can not be achieved. This is summarized in Fig. 10.

For the understanding of the regulation of flavin activities in flavoproteins, as outlined above, we can summarize our mechanistic problems with the flavin system or, in other words, the flavin "region" in the various redox chains by four questions of increasing specificity:

(1) *Redox Stoichiometry*. The question of actual redox stoichiometry being 1 e^- or 2 e^-, as discussed above;

(2) *Nature of Electron Carrier*. In 2 e^--reactions the distinction between H^- (nicotinamide) and R^- (flavin) transfer, as discussed above;

(3) *Geometry of Bonding Interaction*. In the case of R^--transfer we have further to ask for the geometry of the R-Fl-"bond" occurring during catalysis. The easily detectable, colored species often observed caused considerable interest in the last decade, and have been then identified as π-charge transfer complexes[58]. This led to a general tendency to overemphasize π-charge-transfer interactions at the expense of what is called σ-covalent

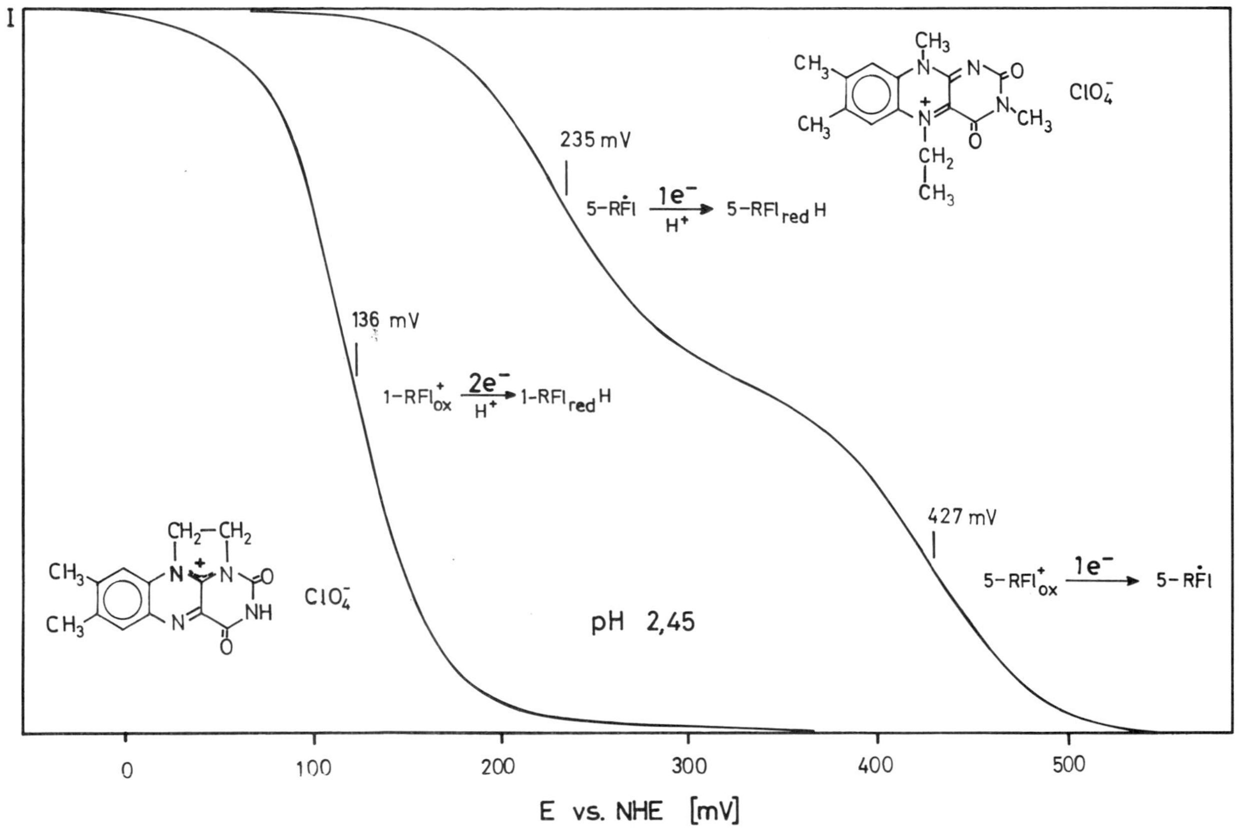

Fig. 7. Polarographic reduction of "5-blocked" versus "1-blocked" flavin[24, 70]. 10^{-4} M solutions of the flavinium perchlorates in 0.1 M sulfate buffer pH 2.45 were reduced polarographically at the rotating platinum electrode. Note the well separated $1\,e^-$-steps in the 5-blocked system as compared to the 1-blocked system. For the corresponding absorption spectra compare Fig. 9[76]. The unsubstituted flavin system behaves under neutral to alkaline conditions as if it was 1-blocked and at $pH < 0$ as if it was 5-blocked. This is due to the fact that the first protonation of the disproportionated system $Fl_{ox} + Fl_{red}H_2$ occurs in the position 5 with concomitant comproportionation of the system ($pK \sim 2.3$)

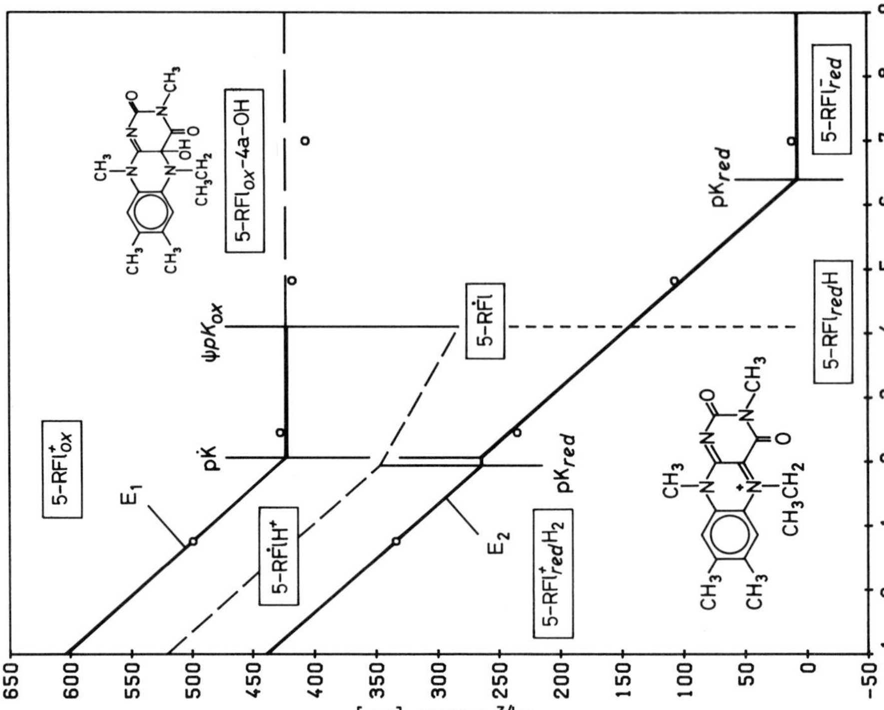
E1/2 vs. NHE [mV]
pH
5-RFl$^+_{ox}$
E1
pK̇
ψpK$_{ox}$
5-RFl$_{ox}$-4a-OH
5-RḞlH$^+$
5-RḞl
E2
pK$_{red}$
5-RFl$^+_{red}$H$_2$
5-RFl$_{red}$H
5-RFl$^-_{red}$
CH$_3$
CH$_3$CH$_2$
OH
N-CH$_3$

Fig. 8b

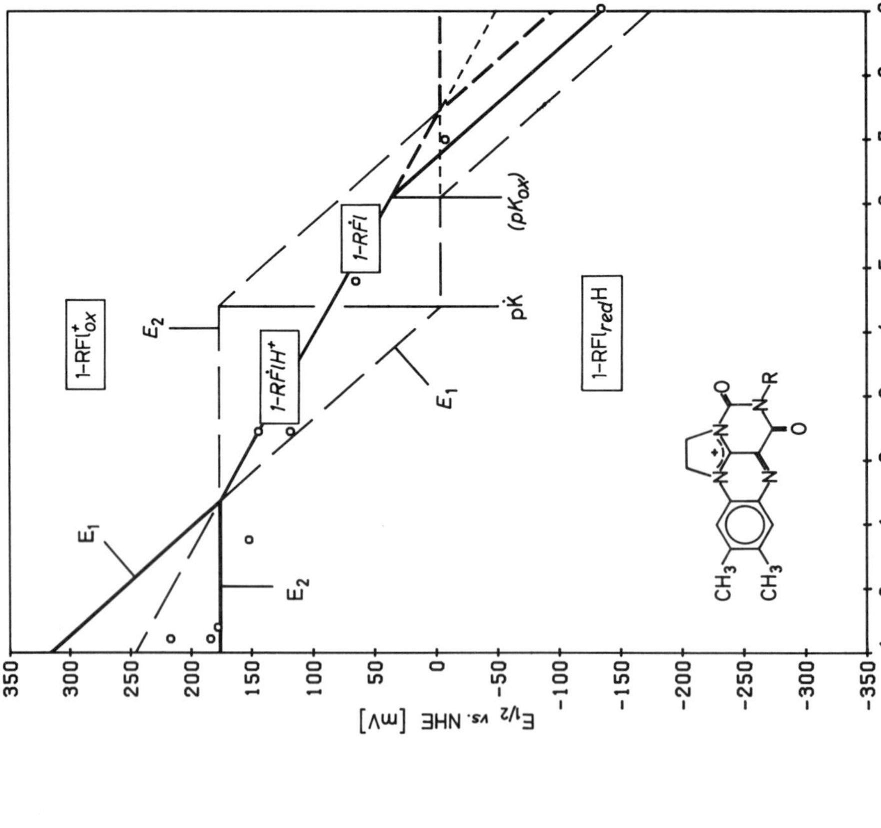
E1/2 vs. NHE [mV]
pH
1-RFl$^+_{ox}$
E1
E2
1-RḞlH$^+$
1-RḞl
pK̇
(pK$_{ox}$)
1-RFl$_{red}$H
CH$_3$
N-R

Fig. 8a

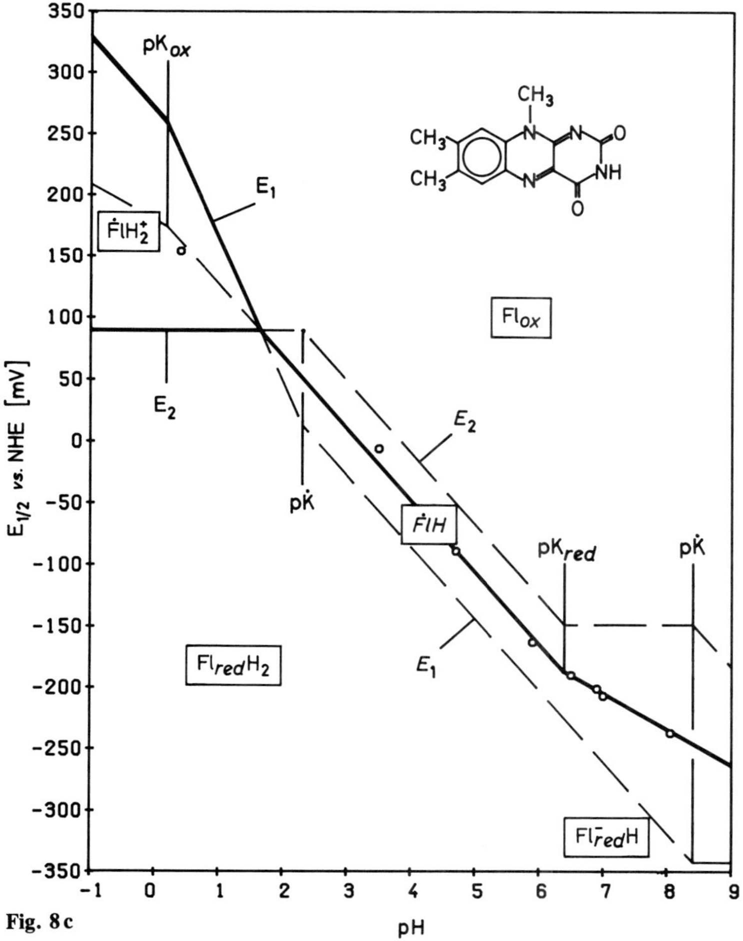

Fig. 8c

Fig. 8 a–c. pH-dependence of redox potentials for **(a)** 1-blocked, **(b)** 5-blocked, **(c)** non-blocked flavins. Whenever at given pH $E_1 > E_2$ there is a preference for single electron transfer, while for $E_2 > E_1$ preference for $2e^-$-transfer is observed. The relation of E_1 versus E_2 is governed by pK-values of the various redox states, each pK presenting a bend in the potential course. The prevailing species depending on potential and pH are indicated by boxes. Note that the 5-blocked as well as to a smaller extent the 1-blocked quinonium salts exhibit a "pseudo pK" i.e. they add hydroxyl ions, instead of liberating protons, which is a slow process, disturbing the thermodynamic accuracy. Experimental conditions (circles indicate measured points): **(a)** and **(b)** polarographic reductions of 0.1 mM 1.10-ethano- resp. 5-ethyl-3-methyl-lumiflaviniumperchlorate in suitable pH-buffer solutions, resp. diluted perchloric or sulfuric acid with the rotating platinum disc electrode[76)]; **(c)** potentiometric titrations of lumiflavin in different acetate or phosphate pH-buffer solutions; in sulfuric acid (pH 0.4) appears the "red intermediate"[49)]

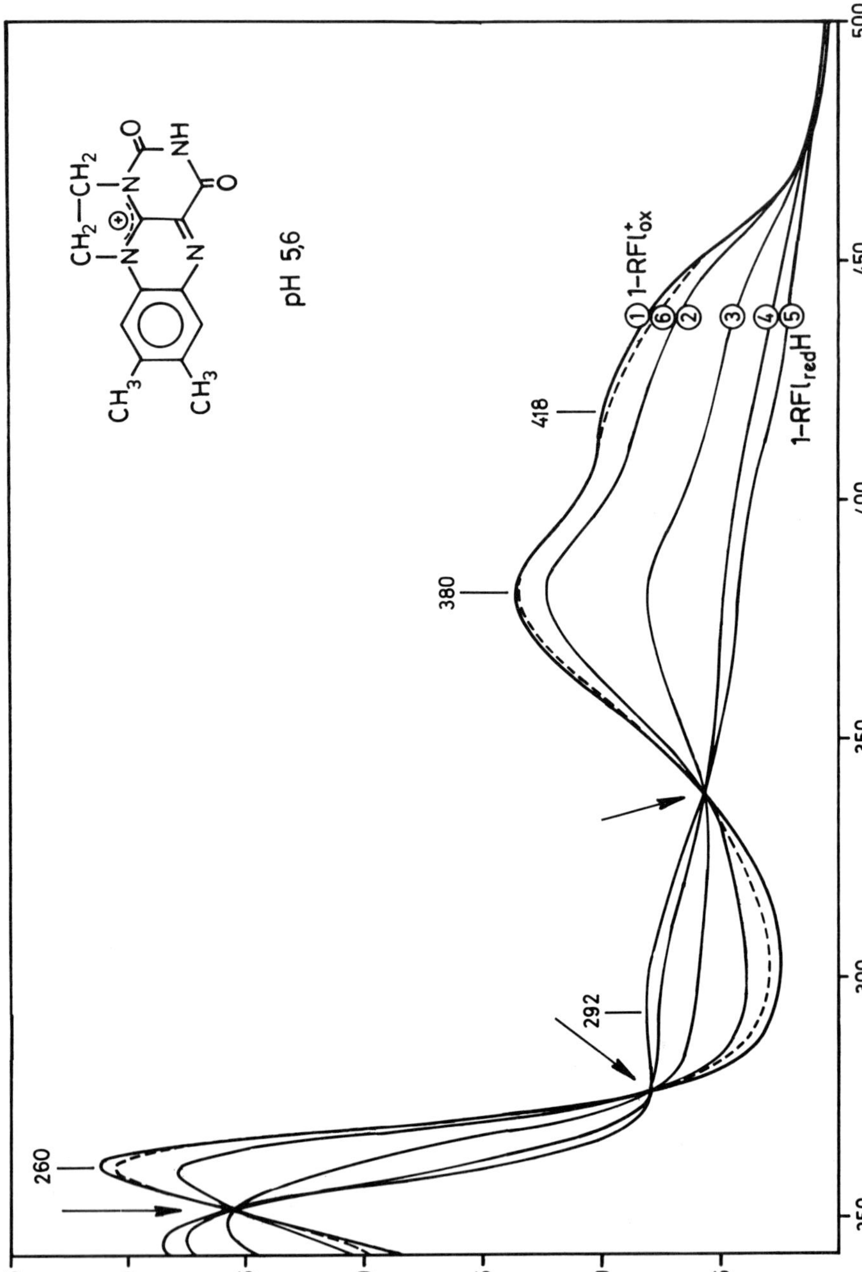
CH$_2$—CH$_2$
CH$_3$
CH$_3$
N
N
N
NH
O
O
pH 5,6
260
292
380
418
① 1-RFl$^+_{ox}$
⑥
②
③
④
⑤ 1-RFl$_{red}$H
ε [mM^{-1}cm^{-1}]
35
30
25
20
15
10
5
250
300
350
400
450
500
λ [nm]

Fig. 9 a

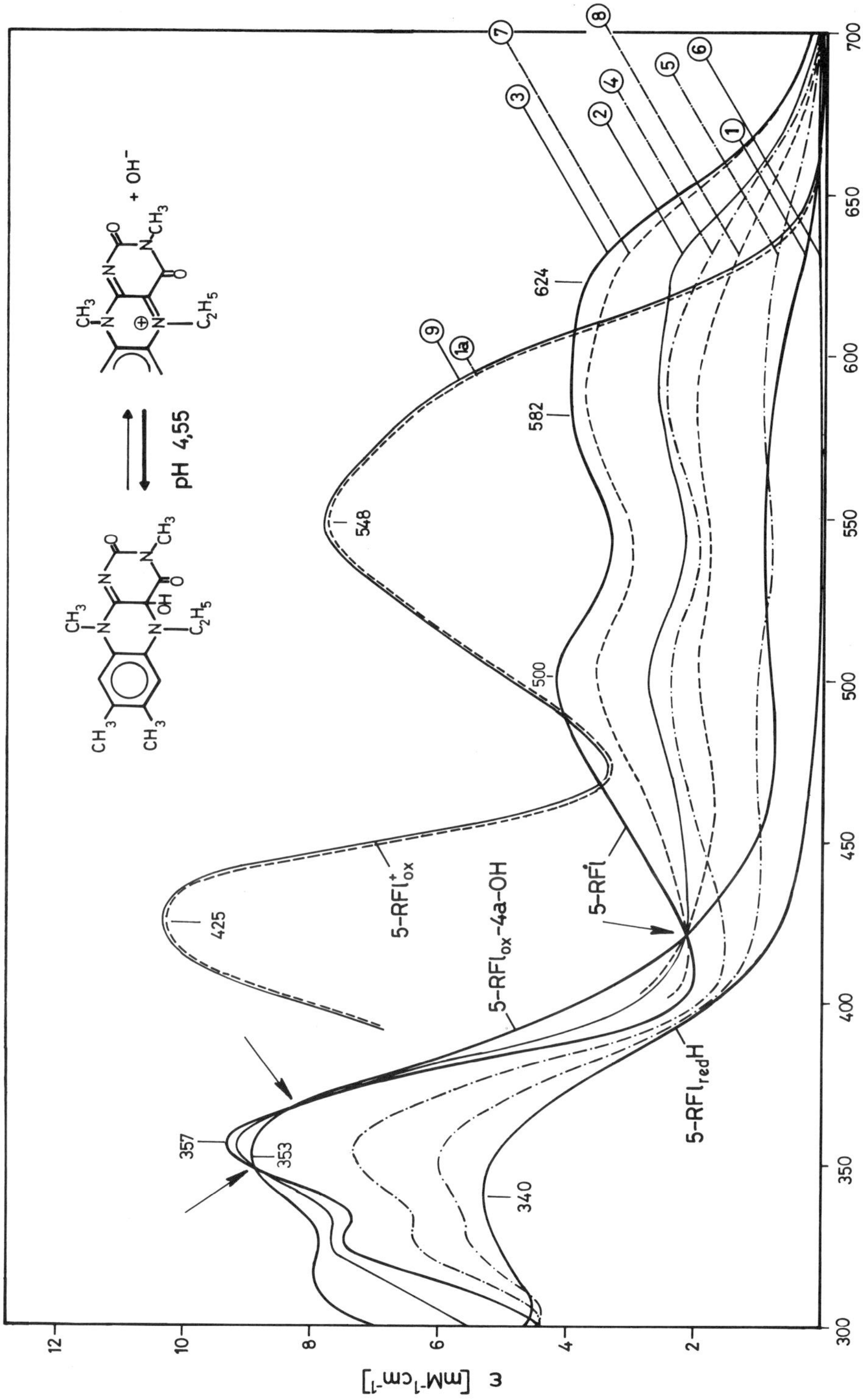

Fig. 9b. (Legend to figure see page 112)

catalysis[36]. Among enzymologists, it is a widespread "mistake" to assume that the σ-character of a substrate coenzyme bond implies kinetic stability. It should be remembered that the prefix σ does not mean this, instead it only denotes an axial bond geometry. In Fig. 5 we showed σ-bonds of increasing kinetic lability for the flavin case. Here we see, that there is a smooth transition from kinetically stable groups such as methyl to alkyl groups migrating on the flavin surface, depending on the pH. At more elevated temperature, such migrating groups become "fluctuating", i.e. they migrate under suitable conditions to and fro so rapidly, that a distinct fixation can not be seen within the measuring time[35, 36]. This, however, does not change the σ-character of the bond. From this it seems to be reasonably clear that the surface of largely planar, aromatic or heteroaromatic biological redox cofactors provides a stabilizing foundation for the migration of organic residues. This is in particular true for the surface of flavocoenzymes[35].

It is a widespread mistake to believe that, maybe because of the larger contact area, π interactions would be more efficient than σ contacts. It has however been shown as visualized in Fig. 11, that in benzoquinhydrone as the prototypic π complex, 1 e^--transfer exhibits an energy barrier, as high as 134 kJ Mol^{-1} and that the two sandwich halves can not approach each other to a distance smaller than 3 Å[82]. This is because the required accompanying proton transfer is hindered, the "intra sandwich" hydrogen bridges not being in the favorable valence angle range of 120 to 180°, but instead at 90°. Orthogonal hydrogen bridges are entirely inefficient and hydrogen bridge formation in flavoquinhydrone will therefore, occur strictly *inter-* and *not intra*molecularly. This proves that an efficient electron transfer in molecular quinhydrone complexes can be excluded under all

---

Fig. 9 a, b. Spectral course of 1-blocked versus 5-blocked flavin reduction. The reductions were performed photochemically[60], in 50% acetonitrile containing 25 mM EDTA buffer of pH 5,6, where EDTA was at the same time photoelectron donor, in the case of 1-RFl_{ox}^{+}, and in 20% acetonitrile containing 8 mM mandelate pH 4,55 in the case of 5-RFl_{ox}^{+}. (**a**) The isosbestic course of the 1-blocked flavin reduction shows that in the neutral pH range no radical is observed at half reduction. ESR-measurements of the half-reduced system (photochemically with mandelate in acetonitrile, which shows optically the identical reduction course as in neutral aqueous solution) gave less than 5% semiquinone[76]. The dashed trace shows practically 100% reversibility of the reduction after admission of O_2. This spectral course is also shown for the flavin dependent reductase in the yeast fatty acid synthetase-complex which is in our definition a pure transhydrogenase (L. Fox, private communication). (**b**) More complicated is the spectral behaviour in the case of the 5-blocked flavin: In this case the starting quinonium salt is only stable at pH < 4. Thus we start with the shown preequilibrium between the quinonium salt absorbing at 548 nm and the nearly colorless 4 a-pseudo base absorbing at 353 nm (*curve 1*). The small amount of cationic chromophore present in solution permits the smooth reduction of the system leading to nearly 100% formation of 5-blocked radical absorbing at 582 to 624 nm (*curve 3*). The intermediate radical is then further reduced to 5-ethyl-$Fl_{red}H$ absorbing at 340 nm (*curve 6*). The reversibility of the system through reduction and reoxidation with oxygen is shown by comparison of the acidified aerobic starting solution (*curve 1 a*), which is identical with the acidified aerobic final solution (*curve 9*): Two sets of isosbestic points are observed, one for the formation and one for the disappearance of the intermediate radical 5-$RFl^{\bullet}$. This spectral course of reduction[76] is identical with what is observed for the reduction of "1 e^- only" flavoproteins such as the flavodoxins[57]

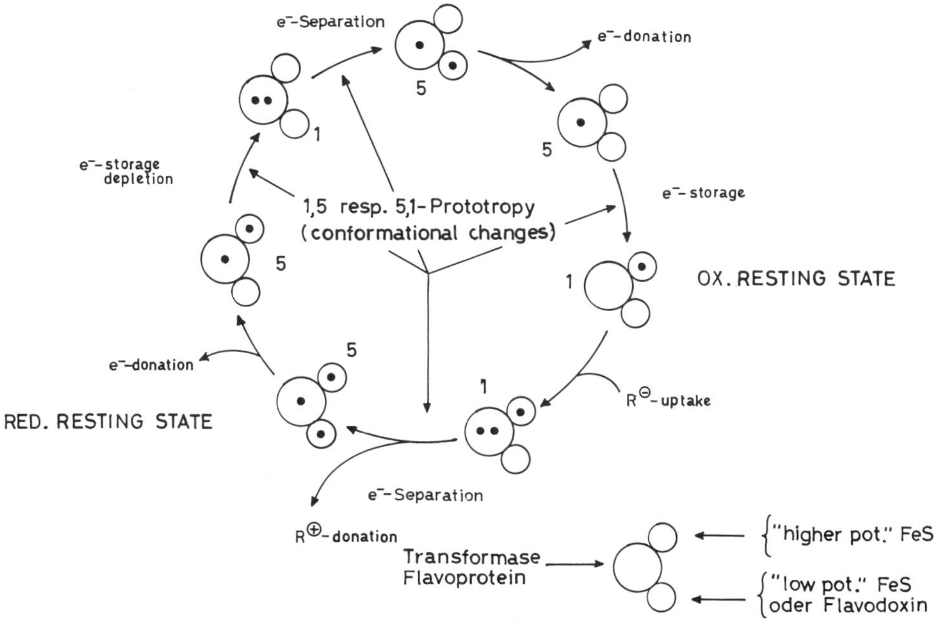

Fig. 10. Flavin-dependent 1 e⁻/2 e⁻-transformase cycle. A *minimum* trinity of redox centers is required consisting of a larger flavin unit and two smaller 1 e⁻-units. The latter may be either FeS centers (photosynthesis and respiration) or one flavodoxin plus FeS centers (azotobacter nitrogen fixation) or a flav(odox)in plus one heme center (P_{450}-oxygenation). The flavin unit can pick up either one or two electrons: In the radical state it is "5-blocked", in the oxidized and fully reduced state 1-blocked and must undergo a conformational change each time it makes a 1 e⁻-step. This is indicated by the arrows stretching out from the center of the cycle. The cycle accounts for the fact that the two steps of 1 e⁻-donation occur from the same 1 e⁻-center with the same potential. Furthermore, RH-dehydrogenation can only proceed, when the 2 e⁻-flavin center is entirely empty. Thus the sequence of events is the following, starting from the "oxidized resting state" which contains one electron in the store: *(1)* R^--uptake in the flavin center followed by leaving of R^+. *(2)* Electron pair splitting and shift of one electron into the low potential FeS-center. By this the "reduced resting state" is reached. *(3)* Donation of the first single electron into the chain followed by return of the store electron into the flavin unit and donation of the second single electron through the low potential FeS center into the chain. Note that the system is never entirely depleted of electrons in the steady state. If there was no store center the cycle would come to a halt whenever it has reached the 1 e⁻-reduced flavin state, since return into the oxidized resting state and further R^--uptake is then prevented

but photochemical conditions, for if the valence angle is rendered favorable we are left with an excessive H-bridge length of > 3.0 Å.

(4) *Site of Substrate Attack.* If a σ-bond is formed, then the next question must pertain to the site of "bonding" of the covalent residue in the flavin. Alkyl fixation at the flavin surface is found chemically in either position 4a, 5, 6 or 8[27,34,35,84,85]. Reversibility of addition as required for biocatalytic relevance, as well as migrational aptitude is encountered in the positions 4a, 5 and 8. In biological systems catalytically competent intermediates have been observed mainly by studies of Ghisla et al.[26,62], all of which have the

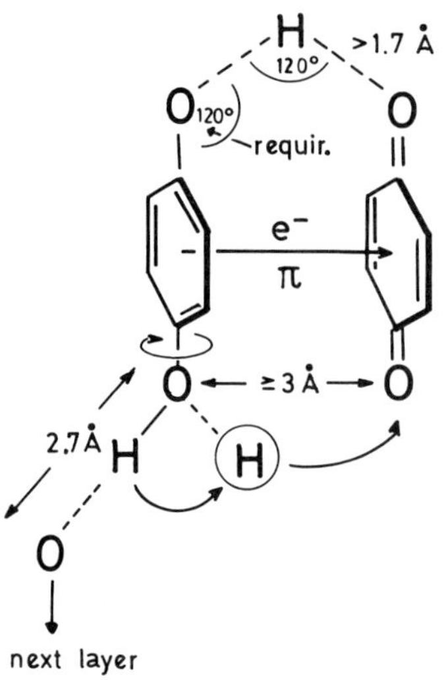

Fig. 11. "Kekulé"-picture of a monomolecular benzoquinhydrone unit in the solid state. Note that the favorable hydrogen bonds are not within this unit but towards the next layer of molecules. This precludes an efficient coupling of electron and proton transfer, and shows that π-interaction of flat donor and acceptor molecules is not by itself an efficient way of catalysis[82]

structure of 5-alkyl dihydroflavins. On the other hand, artificial substitution in the bridge position 4 a, could also be obtained in flavo*proteins,* though not in a reversible way[28] (cf. Eq. 16).

$$Fl_{ox} \begin{array}{l} \nearrow \; 5\text{–R–}Fl_{red}H \\ \qquad\qquad \updownarrows \\ \searrow \; 4a\text{–R–}Fl_{red}H \end{array} \qquad {}^{35)}(16)$$

This failure, however, is easily explained by the fact, that the artificial substituents are too stable, compared with a more labile natural substrate residue. In order to ascertain smooth removal of the alkyl substituent R, this group must bear additional function, such as hydroxyl or amino at the α-carbon[88]. For the lability of such alkyl substituents compare Fig. 5.

Chemically, one can even remove saturated residues R from position 5 via the reaction sequence of Eq. 17:

$$5\text{-RFl}_{red}H \xrightleftharpoons{-e^-/-H^+} 5\text{-R-}\dot{Fl} \xrightleftharpoons{-e^-} 5\text{-R-Fl}^+_{ox} \rightarrow Fl_{ox} + R^+ \qquad (17)$$

The last step will only work smoothly when a suitable alkyl acceptor is present[24].

Summarizing question 4: For biological relevance, we can generalize that flavin-dependent dehydrogenases can fix their substrates by covalent catalysis in position 5 and maybe 4 a[36].

Chemical model experiments let us assume that C(4 a) serves as entrance and leaving position for the reduced R^- whereas N(5) serves the same purpose as for R^+ while intermediate R migration is regulated by pH. A different problem is apparently, the input and output of single electrons in the flavin chromophore. Here we see that the 5-blocked conformation is indeed active (cf. Fig. 6), which points to the fact that the electrons must come and go through the outer edge of flavin nucleus in position 8. This is consistent with the known X-ray data on flavodoxins, as already mentioned[54], and creates a similarity between single electron transfer by flavins and cytochromes. In Fig. 12 we show such an "outer edge contact" between two flavins, in order to demonstrate that even in the electron exchange between such big heteroaromatic planar chromophores, the contact is of σ-geometry, while π-interaction may additionally occur, preferentially in chemical systems, but not as an absolute prerequisite for biological redox catalysis.

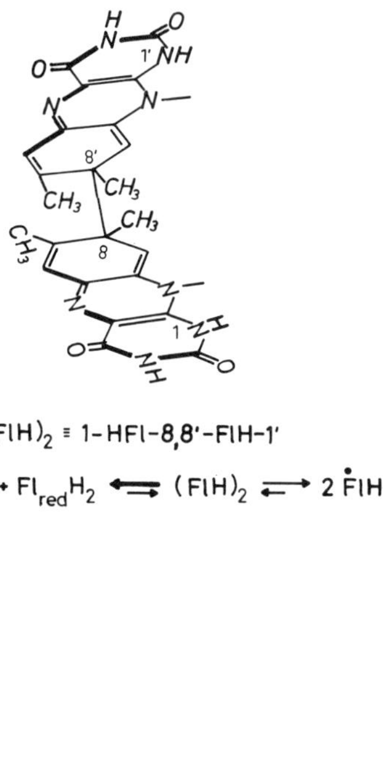

a

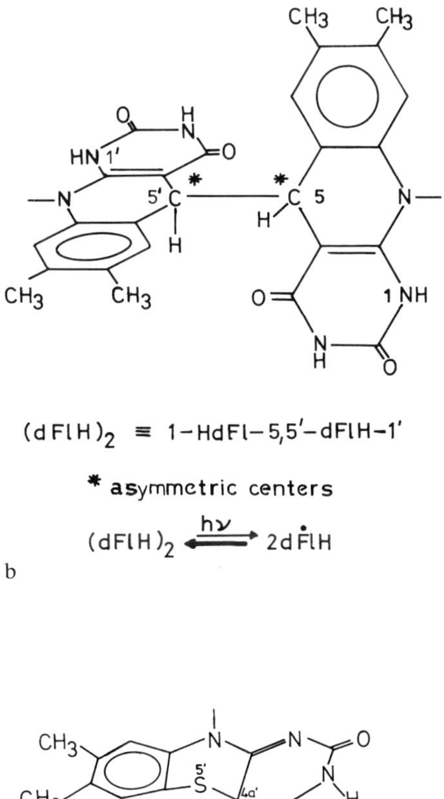

b

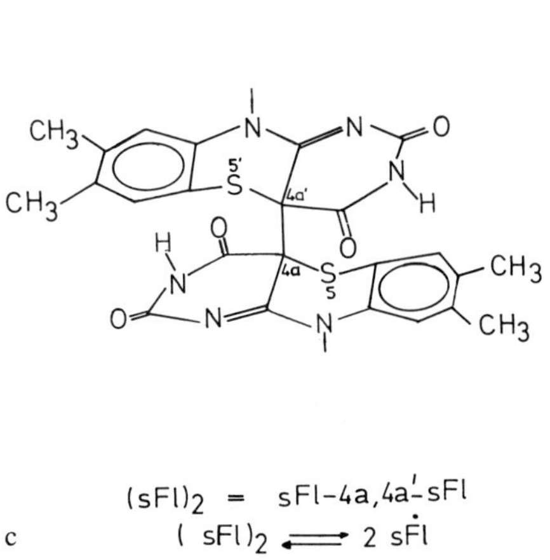

c

Fig. 12 a–c. Structure of σ-covalent half-reduced dimers for **(a)** natural flavin Fl[37], **(b)** 5-deazaflavin dFl[16], and **(c)** 5-thiaflavin sFl[21]. Note that the steric prerequisites are highest for the 5,5′-dimer of dFl and lowest for the 8,8′-dimer of Fl. On the other hand 5,5′-$(dFlH)_2$ is kinetically stable whereas 4a,4a′-$(sFl)_2$ is thermodynamically stable and dissociates thermally into radical monomers. 8,8′-$(FlH)_2$, finally, is kinetically as well as thermodynamically labile and will undergo immediate dismutation. Thus, it is predetermined to be a biocatalytically essential intermediate

N(5) – Mutilated Flavocoenzymes

A further corrobation of the principles outlined above comes from the study of flavocoenzymes modified in the block position N(5) (for structures cf. Fig. 13).

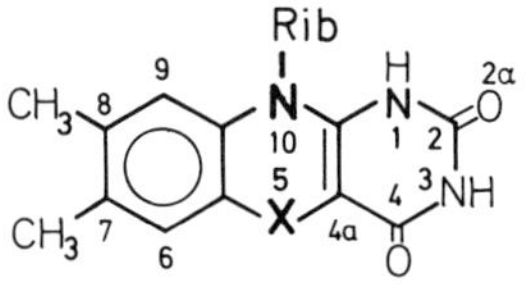

Fig. 13. "Kekulé"-picture of "mutilated flavins" in the fully reduced state. While natural flavin (X = **NH**) shows 3 mechanistically independent activities, namely (de)hydrogenation ($2e^-$), O_2-activation ($1e^-$ or $2e^-$) and single electron transfer ($1e^-$), substitution of X = $\mathbf{CH_2}$ deletes $1e^-$-transfer activity[38], while substitution X = **S** deletes the $2e^-$-transfer activity[21]. This is consistent with $2e^-$-activities being favoured by block of the $1/2\alpha$-lone pairs while $1e^-$-activity is favoured by block of the N(5)-lone pair

In "5-deazaflavin" the flavin center nitrogen 5 is replaced by CH-group. This modification turns the flavin into a "flavin-shaped nicotinamide system"[38], since oxidoreduction now implies formation and rupture of kinetically stable CH- instead of labile NH-bonds.

Hence, 5-deazaflavin dFl_{ox} will react smoothly with chemical $2e^-$-reductants such as borohydride or reduced nicotinamide which is discussed in a preceding section (cf. Eq. 3c). It should be noted that natural flavin reacts slowly with these $2e^-$-reductants (cf. Eq. 4c). If, however, the uptake of single electrons in the 5-deazaflavin system is enforced by method of Massey and Hemmerich[60] (EDTA in the presence of light), the first intermediate to be observed is the red radical 1-HdFl• (cf. Eq. 3a), which is very short lived and is a strong reductant (cf. Table 1)[29]. This radical in turn will back donate its single electron smoothly to any suitable $1e^-$-acceptor in the environment. This is the reason why this method lends itself easily for the controlled reduction of many sensitive and vulnerable redox proteins.

In order to study enforced $1e^-$-transfer in a $2e^-$-specific redox system, 5-deazaflavin offers the following advantages over nicotinamide:

The formation of the primary radical NAD• leads to the dead end of irreversible dimerization[9]. The primary radical 1-HdFl• formed in the $1e^-$-reduction of deazaflavin, however, decays preferably by dismutation[29]. This dismutation is photoreversible, as in general photochemically active semiquinone dismutation equilibria are reversed by the action of light (cf. Eq. 18a; for the corresponding dismutation equilibrium for flavins, cf. Eq. 13).

$$2\ \ 1\text{–Hd}\dot{\text{F}}\text{l} \rightleftharpoons \text{dFl}_{ox} + \text{H}_2\text{dFl}_{red} \quad \text{a}$$

$$2\ \ 1\text{–Hd}\dot{\text{F}}\text{l} \overset{slow}{\rightleftharpoons} 1\text{–HdFl–5, 5'–dFlH–1'} \quad \text{b} \qquad (18)$$

Hence, under conditions of long-term illumination, a second and slower decay channel for 1-HdFl• comes to bear, namely 5,5′-dimerization[16] (cf. Fig. 12 b). This reaction is slower because the 5,5′-contact is sterically somewhat inhibited. It must, therefore, be concluded, that the rapid photoreversible dismutation apparently does not proceed through a slow, but stable 5,5′- but instead (as already hypothesized above) a fast, but labile 8,8′-contact (cf. Fig. 12 a). Under continuous illumination the slowly formed 5,5′-dimer $(HdFl)_2$ (cf. Eq. 18 b) will, therefore, be the only end product of deazaflavin photoreduction. In contrast to $(NAD)_2$ (cf. Eq. 5) the deazaflavin dimer can, however, be reconverted to oxidized starting material by mild oxidation or photooxidation[16, 29].

It should be further noted, that in the deazaflavin photoreduction, part of the light energy is stored, owing to the formation of the stable dimer 5,5′-$(HdFl)_2$, while in the half-reduced natural flavin system, the energy gained by photocomproportionation is dissipated during the subsequent dark re-dismutation[40] (cf. Eq. 13 and Table 2).

On the other hand, the differences between the 5-deazaflavin and the natural flavin system lie in the following facts:

1) We saw that the flavin system exhibits two types of radicals which are tautomers, namely the red 1/2 α-HFl•, which is under all chemical conditions thermodynamically and kinetically unstable, and the blue 5-HFl•, which can be stabilized chemically by alkylation and biologically by insertion into a protein. Interconversion of these radicals must be rapid, since it invokes only shift of mobile hydrogen. Every flavin-dependent biological 1 e^-/2 e^--transformation must be accompanied as mentioned by such rapid prototropically induced conformation change.
2) In the 5-deazaflavin system, however, the interconversion of the tautomeric 1- and 5-blocked radicals is slowed down, because it requires formation and rupture of C(5)-H bonds (cf. Fig. 14). Thus, we obtain, as mentioned, the 1-blocked red and strongly reducing radical upon 1 e^--reduction of the quinone dFl_{ox}, while the primary product of the 1 e^--oxidation of hydroquinone H_2dFl_{red} is the colorless and oxidizing radical 5-HdFl[29]. In contrast to its tautomer this radical can not undergo direct dismutation nor formation of a stable dimer and must, therefore, await its slow 5-deprotonation, if left to itself.

In summary the biologically required prototropic interconversion of 1/2 α- and 5-blocked radical tautomers, which is fast in the free flavin system, is slowed down in the 5-deazaflavin analog. This modification, thus, provides us with a model, in which the conformational shuttle accompanying the 1 e^-/2 e^--transformation is slowed down. Thus, we arrive at a model of a biological "1 e^-/2 e^--transformase" machine, switched down to low gear.

From this it follows further that the 5-deazaflavin modification deletes specifically two of the three main flavin activities, namely single electron transfer and dioxygen activation, while it transforms the natural flavin-dependent dehydrogenation activity, which is a group transfer, into a nicotinamide-simulating hydride transfer[38].

Finally, we can also demonstrate the reverse principle, when replacing the center nitrogen in natural flavin by sulfur: Thus, we arrive at a "5-thiaflavin" system, which still lends itself for the abstraction and back donation of single electrons with formation of stable radicals sFl•. This radical is trivially 5-blocked, having no mobile proton and thus can not undergo dismutation, but only dimerization in the sterically less favorable positions 4 a,4 a′ [21](cf. Eq. 19). This σ-dimer formation (cf. Fig. 12 c) can even be reversed thermally, though the dimer 4 a,4 a′-$(sFl)_2$ can be isolated easily at low temperature.

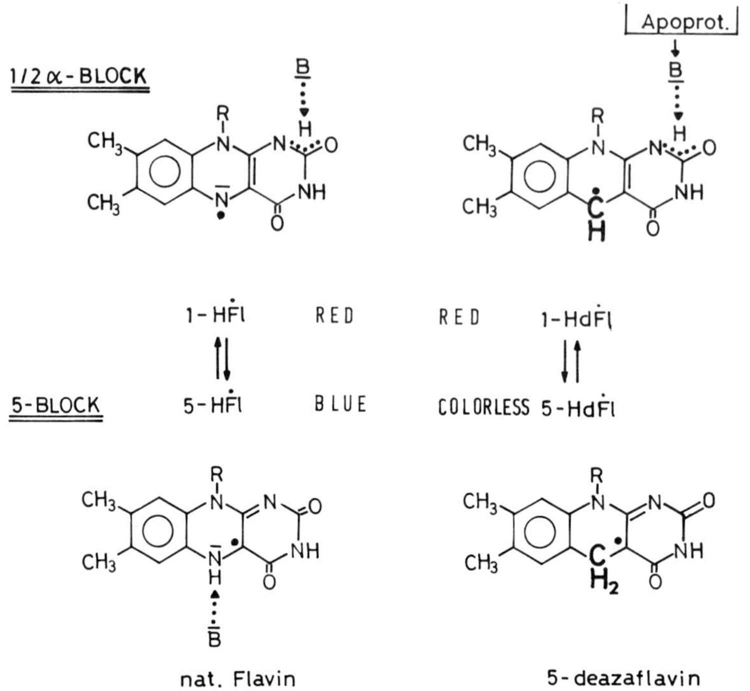

Fig. 14. Supplementation of Fig. 13 showing for natural flavin and 5-deazaflavin the possible neutral radical tautomers which in the 1/2 α-blocked case are red-coloured and thermodynamically unstable[18, 72], whereas in the 5-blocked case the radical is blue and thermodynamically stable for natural flavin[72] and colorless and unstable for deazaflavin[29]. Note that the interconversion as well as the dismutation and comproportionation of these radicals is fast for natural flavin but slow for deazaflavin since in this latter case CH-bonds must be formed and split. On the other hand the natural radical prototropism can be regulated (i.e. accelerated as well as slowed down) by the lone pair blocking action of the apoprotein

$$HsFl_{red} \xrightarrow[+430\ mV]{-e^-/-H^+} sFl^{\bullet} \rightarrow \tfrac{1}{2}\ sFl\text{-}4a{,}4a'\text{-}sFl \qquad (19)$$

The sulfur modification eliminates two among the three flavin activities, namely (de)hydrogenation and oxygen activation. The remaining slow oxygen affinity of 5-thiaflavin is a $1e^-$-transfer affinity, yielding superoxide as first intermediate (cf. Eq. 11). The same affinity is characteristic as an artificial side reaction in flavin-dependent dehydrogenases, leading to the formation of superoxide as a biological accident, which could be repaired by the interaction of the ever-present superoxide dismutase.

Summarizing, the modification in the flavin N(5) position renders it possible to create "mutilated" flavins and flavoproteins, which in turn give us entirely new insights into flavin-dependent redox transfer mechanisms, and allow us to separate $1e^-$- and $2e^-$-transfer capacities, both of which are inherent in the natural flavin system, but regulated by apoprotein.

Final Remarks

The scope and limitation of single electron transfer in biology is still a matter of controversy. In particular the uniqueness of flavin as $1\,e^-/2\,e^-$-interconverting agent requires further examination after the description of a steadily increasing number of molybdenum-containing enzyme systems which appear to shuttle between the valence states of IV and VI in a way which is general $2\,e^-$, but apparently leaves the opportunity of "electron pair splitting" under certain conditions which are still ill-defined[90]. The ESR spectrum of the Mo(V) state is seen during the reoxidation of these systems, but quantitative determination of the single electron step is difficult because of the complicated dismutation equilibrium (cf. Eq. 20).

$$2\,\mathrm{Mo(V)} \leftrightharpoons \mathrm{Mo(IV)} + \mathrm{Mo(VI)} \qquad (20)$$

This represents in turn the uniqueness of molybdenum as a "carrier free $2\,e^-$-transferring agent"[91], that is in other words, equivalent to the designation of Mo as a "$2\,e^-$-store" which is rapidly accessible and available in contrast to, say, NADH as a kinetically stable hydride store.

In this context, one is quite generally referred to the question of *thermodynamic versus kinetic* radical stability: Unless particular pathways for a "dismutational" $1\,e^-$-exchange between redox centers are foreseen by nature, one would assume that dismutation of protein molecules is always slow. This kinetic problem adds to the thermodynamic problem, and in every treatment of a flavoprotein radical both aspects have to be considered.

There remains the question whether "everything is possible in redox enzymology" or whether certain pathways of redox equivalent transfer are definitely to be excluded. In the following sentences we want to propose various restrictions of general relevance:

1) No flow of redox equivalents is possible between nicotinamide and any $1\,e^-$-center, whether it be heme or iron sulfur, without the intermediate action of a flavin.
2) The difference between nicotinamide, flavin- and molybdenum-dependent dehydrogenations consists in the nature of the substrate bond to be broken in the first activation step. For nicotinamide as mentioned, this is $C^{\delta+}(sp^3)H^{\delta-}$, for flavin it is $C^{\delta-}(sp^3)H^{\delta+}$ and for molybdenum it is $C^{\delta+}(sp^2)H^{\delta-}$. Hence we want to postulate the restriction that dehydrogenation of trigonal CH has an absolute requirement for molybdenum as acceptor.
3) A special unresolved problem is the question of flavin being essential or not in the (de)hydrogenation of α,β-unsaturated carboxylic acid derivatives as it occurs during biosynthesis and degradation of fatty acids. Little emphasis has been paid up to now by investigators in this field, to the fact that depending on the nature of the organism, flavin is required in some cases for fatty acid synthesis such as in yeast, but is apparently not required in e.g. mammalian liver. This would be a paradigmatic case of "everything may happen in proteins (or not)". Hence the problem is posed whether nicotinamide could reduce such systems. Whether or not a 1,4-hydride transfer from nicotinamide to an α,β-unsaturated acid is possible at all, it must be said to be not only unresolved but largely unconsidered.

It is a prediction of this review that in principle nicotinamide cannot replace flavin and vice versa, not even within a range of $2e^-$-transfer reactions, because nicotinamide transfers hydride while flavin deals with carbanionic substrate residues.
On the other hand a probability not to be neglected is the following one: Flavin is often the limiting factor in many cases of multifactor redox systems. This is apparently true for the mitochondrial respiratory chain.
On the other hand it is a wide spread but unjustified belief that flavocoenzymes bind their cofactor so vigorously that it can not come off without causing irreversible damage. In contrast flavin may not only be limiting but it may be the cofactor most easily lost in a given system, and precisely those flavin systems where the flavin is not bound too tightly will elapse speedy recognition to the extent that the whole system might be rate determined by substoichiometric amounts of flavin which have not been washed out during preparation. In such cases one should always try reconstitution experiments with extraneous FMN and/or FAD.
A pertinent question in this matter concerns the structural requirement for substrates of flavin dependent dehydrogenation of α,β-unsaturated acid derivatives. From all we know it appears that carboxylate anions can not be substrates of flavin dependent dehydrogenases without being activated
- either by conversion to an activated ester, such as an acyl-CoA derivative,
- or by coordination in an iron-sulfur-cluster belonging to the apoenzyme.

This case appears to be verified in the normal succinate dehydrogenases, as well as in microbial enoate-reductase[92].
In any case the first step in the dehydrogenation of the substrate is α-deprotonation and this step must be facilitated by preceding suitable derivatisation of the carboxylate.

4) Hence, a ready explanation for the fact is at hand, that flavin dependent carboxylate dehydrogenation finds partly activated substrates and partly inactivated ones. On the other hand we know a number of flavin substrates whose CH-bonds are split without activation by a further electron withdrawing substituent. Examples to be mentioned here are the flavin dependent alcohol oxidases such as methanol oxidase or glycerophosphate oxidase. These enzymes catalyze dehydrogenation of simple alcoholic substrates, and it is difficult to conceive that the apoprotein should be able in these cases to afford the high activation energy required for CH-deprotonation of a simple alcohol. We postulate that redox transfer is carbanionic also in this last mentioned case, because it is flavin and not nicotinamide-dependent. We must at the same time think how the apoenzyme could be of help for this purpose. Some help for better understanding may come from the fact that replacement of the $2e^-$-substrate methanol by the $1e^-$-analog hydroxylamine switches the enzyme from $1e^-$- towards $2e^-$-transfer, and consequently radical stabilization[23]. The radical observed here is the red 1-blocked semiquinone anion. It is so stable that it cannot be discharged by procedures, which leave the protein intact.

References

1. Biellmann, C. F., Jung, M. J.: FEBS-Letters *7,* 199 (1970)
2. Blankenhorn, G.: Eur. J. Biochem. *50,* 351 (1975)
3. Blankenhorn, G.: Eur. J. Biochem. *67,* 67 (1976)
4. Blankenhorn, G., Moore, E. G.: J. Am. Chem. Soc. *102,* 1092 (1980)
5. Bors, W., Michel, C., Saran, M.: Eur. J. Biochem. *95,* 621 (1979)
6. Bright, H. J., Porter, D. J. T.: In: The Enzymes, Vol. XII B. P. D. Boyer (ed.), p. 421, New York: Academic Press 1975
7. Brown, H. C., Krishnamurthy, S.: Tetrahedron *35,* 567 (1979)
8. Buddecke, E.: Grundriß der Biochemie, 4th ed., p. 44, Berlin: de Gruyter 1974
9. Burnett, R. W. Underwood, A. L.: Biochemistry *7,* 3328 (1968)
10. Caughey, W. S., Choc, M. G., Houtchens, R. A.: In: Biochemical and Clinical Aspects of Oxygen. W. S. Caughey (ed.), p. 1, New York: Academic Press 1979
11. Cava, M. P., Mitchell, M. J.: Cyclobutadiene and Related Compounds, New York: Academic Press 1967
12. Chen, S. J., Fowler, F. W.: J. Org. Chem. *35,* 3987 (1970)
13. Chen, S. J., Fowler, F. W.: J. Org. Chem. *36,* 4025 (1971)
14. Dixon, M., Webb, E. L.: Enzymes, 3rd ed., p. 474, London: Longmann 1979
15. Draganic, I. C., Draganic, Z. D.: The Radiation Chemistry of Water, p. 70, New York – London: Academic Press 1971
16. Duchstein, H.-J., Fenner, H., Hemmerich, P., Knappe, W.-R.: Eur. J. Biochem. *95,* 167 (1979)
17. Dudley, K. H., Ehrenberg, A., Hemmerich, P., Müller, F.: Helv. Chim. Acta *47,* 1354 (1964)
18. Ehrenberg, A., Müller, F., Hemmerich, P.: Eur. J. Biochem. *2,* 286 (1967)
19. Eisner, U., Kuthan, J.: Chem. Rev. *72,* 1 (1972)
20. Fee, J. A.: Struct. Bonding (Berlin) *23,* 1 (1975)
21. Fenner, H., Grauert, R., Hemmerich, P., Michel, H., Massey, V.: Eur. J. Biochem. *95,* 183 (1979)
22. Frisell, W. R., Chung, C. W., Mackenzie, C. G.: J. Biol. Chem. *234,* 1297 (1959)
23. Geißler, J., Hemmerich, P.: FEBS-Letters, *126,* 152 (1981)
24. Ghisla, S., Hartmann, U., Hemmerich, P., Müller, F.: Liebigs Ann. Chem. *1973,* 1388
25. Ghisla, S., Massey, V., Lhoste, J.-M., Mayhew, S. G.: Biochemistry *13,* 589 (1974)
26. Ghisla, S., Massey, V.: In: Mechanisms of Oxidizing Enzymes. T. P. Singer, R. Ondarza (ed.), p. 55, Amsterdam: Elsevier 1978
27. Ghisla, S., Olson, S. T., Massey, V., Lhoste, J.-M.: Biochemistry *18,* 4724 (1979)
28. Ghisla, S., Wenz, A., Thorpe, C.: In: Enzyme Inhibitors. U. Brodbeck (ed.), p. 43, Weinheim: Verlag Chemie 1980
29. Goldberg, M., Pecht, I., Kramer, H. E. A., Traber, R., Hemmerich, P.: Biochim. Biophys. Acta, *673,* 570 (1981)
30. Hayon, E., Simic, M.: Acc. Chem. Res. *7,* 114 (1974)
31. Hemmerich, P., Prijs, B., Erlenmeyer, H.: Helv. Chim. Acta *43,* 372 (1960)
32. Hemmerich, P., Nagelschneider, G., Veeger, C.: FEBS-Letters *8,* 69 (1970)
33. Hemmerich, P., Schuman-Jorns, M.: In: Enzymes: Structure and Function. J. Drenth, R. A. Oosterbaan, C. Veeger (ed.), p. 95, Amsterdam: North-Holland Pub. Corp. 1972
34. Hemmerich, P., Bhaduri, A. P., Blankenhorn, G., Brüstlein, M., Haas, W., Knappe, W.-R.: In: Oxidases and Related Redox Systems. T. E. King, H. S. Mason, M. Morrison (ed.), p. 3, Baltimore: University Park Press 1973
35. Hemmerich, P., Haas, W.: In: Reactivity of Flavins. K. Yagi (ed.), p. 1, Tokyo: University Press, 1975
36. Hemmerich, P.: In: Progress in the Chemistry of Organic Natural Products. W. Herz, H. Grisebach, G. W. Kirby (ed.), *33,* 451 (1976)
37. Hemmerich, P.: Adv. Chem. Ser. *162,* 312 (1977)
38. Hemmerich, P., Massey, V.; Fenner, H.: FEBS-Letters *84,* 5 (1977)
39. Hemmerich, P., Wessiak, A.: In: Biochemical and Clinical Aspects of Oxygen. W. S. Caughey (ed.), p. 491, New York: Academic Press 1979

40. Hemmerich, P., Knappe, W.-R., Kramer, H. E. A., Traber, R.: Eur. J. Biochem. *104,* 511 (1980)
41. Hemmerich, P., Massey, V.: In: Oxidases and Related Redox Systems. T. E. King, H. S. Mason, M. Morrison (ed.) in press 1981
42. Hutton, R. F., Westheimer, F. H.: Tetrahedron *3,* 73 (1958)
43. Jagow, G. v., Engel, W. D.: Angew. Chem. *92,* 684 (1980)
44. Jefcoate, C. R., Ghisla, S., Hemmerich, P.: J. Chem. Soc. (C) *1971,* 1689
45. Jonah, C. D., Miller, J. R., Matheson, M. S.: J. Phys. Chem. *81,* 1618 (1977)
46. Karlson, P.: Kurzes Lehrbuch der Biochemie für Mediziner und Naturwissenschaftler, 8th ed., p. 89, Stuttgart: Thieme Verlag 1972
47. Kierkegaard, P., Norrestam, R., Werner, P., Csöregh, I., Glehn, M. v., Karlsson, R., Leijonmarck, M., Rönnquist, O., Stennesland, B., Tillberg, O., Torbjörnsson, L.: In: Flavins and Flavoproteins. H. Kamin (ed.), p. 1, Baltimore: University Park Press 1971
48. Knappe, W.-R., Hemmerich, P.: Liebigs Ann. Chem. *1976,* 2037
49. Kuhn, R., Boulanger, P.: Ber. *69,* 1557 (1936)
50. Kuhn, R., Ströbele, R.: Ber. *70,* 753 (1937)
51. Lambeth, D. O., Palmer, G.: J. Biol. Chem. *248,* 6095 (1973)
52. Land, E. J., Swallow, A. J.: Biochemistry *8,* 2117 (1969)
53. Lehninger, A. L.: Biochemie, 2nd ed., p. 276, 399, Weinheim: Verlag Chemie 1977
54. Ludwig, M. L., Burnett, R. M., Darling, C. D., Jordan, S. R., Kendall, D. S., Smith, W. W.: In: Flavins and Flavoproteins. T. P. Singer (ed.), p. 393, Amsterdam – Oxford – New York: Elsevier: North-Holland 1976
55. Mahler, H. R., Cordes, E. H.: Biological Chemistry, 2nd ed., p. 413, New York: Harper & Row 1971
56. Malmström, B. G., Andrèasson, L.-E., Reinhammar, B.: In: The Enzymes, Vol. XII B. P. D. Boyer (ed.), p. 507, New York: Academic Press 1975
57. Massey, V., Palmer, G.: Biochemistry *5,* 3181 (1966)
58. Massey, V., Ghisla, S.: Ann. N. Y. Acad. Sci. *227,* 446 (1974)
59. Massey, V., Hemmerich, P.: In: The Enzymes, Vol. XII, 3rd ed., P. D. Boyer (ed.), p. 191, New York: Academic Press 1975
60. Massey, V., Hemmerich, P.: Biochemistry *17,* 9 (1978)
61. Massey, V.: In: Biochemical and Clinical Aspects of Oxygen. W. S. Caughey (ed.), p. 477, New York: Academic Press 1979
62. Massey, V., Ghisla, S., Kieschke, K.: J. Biol. Chem. *255,* 2796 (1980)
63. Massey, V., Hemmerich, P.: Biochem. Soc. Trans. Biochem. Rev. *8,* 246 (1980)
64. Mayhew, S. G., Ludwig, M. L.: In: The Enzymes, Vol. XII B, 3rd ed. P. D. Boyer (ed.), p. 57, New York: Academic Press 1975
65. Mayhew, S. G., van Arem, E. J. F., Strating, M. J. J., Wassink, J. H.: In: Flavins and Flavoproteins. T. D. Singer (ed.), p. 411, Amsterdam – Oxford – New York: Elsevier, North-Holland 1976
66. Metzler, D. E: Biochemistry, p. 464, New York: Academic Press 1977
67. Michaelis, L., Schubert, P. M., Smythe, C. V.: J. Biol. Chem. *116,* 587 (1936)
68. Michaelis, L., Schwarzenbach, G. J.: J.Biol. Chem. *123,* 527 (1938)
69. Mincey, T., Tayrien, G., Mildvan, A. S., Abeles, R. H.: Proc. Natl. Acad. Sci. USA *77,* 7099 (1980)
70. Müller, F., Massey, V.: J. Biol. Chem. *244,* 4007 (1969)
71. Müller, F., Massey, V., Heizmann, C., Hemmerich, P., Lhoste, J.-M., Gould, D. C.: Eur. J. Biochem. *9,* 392 (1969)
72. Müller, F., Brüstlein, M., Hemmerich, P., Massey, V., Walker, W.: Eur. J. Biochem. *25,* 573 (1972)
73. Palmer, G., Müller, F., Massey, V.: In: Flavins and Flavoproteins. H. Kamin (ed.), p. 123, Baltimore: University Park Press 1971
74. Pfleiderer, W., Mengel, R., Hemmerich, P.: Chem. Ber. *104,* 2273 (1971)
75. Schmidt, R. R.: Angew. Chem. *87,* 603 (1975)
76. Schug, C.: Diploma Thesis, University of Constance (1978)
77. Stewart, R., Norris, D. J.: J. C. S. Perkin II, *1978,* 246
78. Stryer, L.: Biochemie, p. 180, Braunschweig: Vieweg 1979

79. Tauscher, L., Ghisla, S., Hemmerich, P.: Helv. Chim. Acta, *56,* 630 (1973)
80. Vaish, S. P., Tollin, G.: Bioenergetics *2,* 61 (1971)
81. Van Dijk, C., Veeger, C.: Eur. J. Biochem. *114,* 209 (1981)
82. Vinogradov, S. N., Linnell, R. H.: Hydrogen Bonding. New York: Van Nostrand Reinhold Company 1971
83. Visser, A. J. W. G., Müller, F.: Helv. Chim. Acta *62,* 593 (1979)
84. Walker, W., Hemmerich, P., Massey, V.: Helv. Chim. Acta *50,* 2269 (1967)
85. Walker, W., Hemmerich, P., Massey, V.: Eur. J. Biochem. *621,* 166 (1959)
86. Wallenfels, K., Diekmann, K.: Liebigs Ann. Chem. *621,* 166 (1959)
87. Walsh, C.: Enzymatic Reaction Mechanisms, p. 427 ff. San Fransisco: W. H. Freeman 1979
88. Walsh, C.: Acc, Chem. Res. *13,* 148 (1980)
89. Wieland, H.: J. Chem. Soc. *1931,* 1055
90. Molybdenum and Molybdenum-Containing Enzymes, Coughlan, M. (ed.), Oxford–New York: Pergamon Press 1980
91. Olson, J. S., Ballou, D. P., Palmer, G. and Massey, V.: J. Biol. Chem. *249,* 4363 (1974)
92. Tischer, W., Bühler, M., Simon, H. and Hemmerich, P.: Z. Physiol. Chem. *361,* 339 (1980)

Author-Index Volumes 1–48

G. Habermehl

Venomous Animals and Their Toxins

Based on the translation of the 2nd German Edition of "Gift-Tiere und ihre Waffen"

1981. 44 figures, 44 tables. Approx. 150 pages
ISBN 3-540-10780-0

Contents: Introduction. – Protection from Snake Bites. – Coenlenterates and Cnidaria. – Mollusca (Molluscs). – Athropoda. – Echinodermata (echinoderms). – Pisces (fishes). – Amphibia (amphibians). – Reptilia (reptiles). – Therapeutic Use of Animal Venoms. – Subject Index. – List of Institutes Which Produce Antivenins.

Venomous Animals and Their Toxins is a unique presentation of our knowledge of the venoms of marine and terrestrial animals who use their "weapons" for hunting and self-defence. The author, President of the European Section of the International Society of Toxicology, has himself done a considerable amount of chemical research in this area, and has been especially active in the elucidations of the structure of salamander venoms.

Although directed primarily to biologists and chemists, this book is also a source of valuable information for tourists who may come into contact with venomous animals. The purpose of the book is not, however, to try and convince the reader that danger from venomous animals lurks around every corner; instead, it shows how to avoid accidents with these animals and how to behave in case one should occur. With its new insights into the modes of action and constituents of different venoms, and into the significance of the venoms for human beings and for clinical application, *Venomous Animals and Their Toxins* will also prove to be of interest to physicians and medical researchers.

Springer-Verlag
Berlin
Heidelberg
New York

A. F. Williams

A Theoretical Approach to Inorganic Chemistry

1979. 144 figures, 17 tables. XII, 316 pages
ISBN 3-540-09073-8

Contents: Quantum Mechanics and Atomic Theory. – Simple Molecular Orbital Theory. – Structural Applications of Molecular Orbital Theory. – Electronic Spectra and Magnetic Properties of Inorganic Compounds. – Alternative Methods and Concepts. – Mechanism and Reactivity. – Descriptive Chemistry. –Physical and Spectroscopic Methods. – Appendices. – Subject Index.

This book outlines the application of simple quantum mechanics to the study of inorganic chemistry, and shows its potential for systematizing and understanding the structure, physical properties, and reactivities of inorganic compounds. The considerable strides made in inorganic chemistry in recent years necessitate the establishment of a theoretical framework if the student is to acquire a sound knowledge of the subject. A wide range of topics is covered, and the reader is encouraged to look for further extensions of the theories discussed. The book emphasizes the importance of the critical application of theory and, although it is chiefly concerned with molecular orbital theory, other approaches are discussed. This text is intended for students in the latter half of their undergraduate studies. (235 references)

Springer-Verlag
Berlin
Heidelberg
New York